Natural Science in Archaeology

The last three decades have seen a steady growth of application of natural scientific methods to archaeology. The interdisciplinary approach of archaeometry has found increasing appreciation by the archaeologists and is now considered indispensable and an integral part of archaeological studies. Interdisciplinary collaboration requires a multidisciplinary background. It is becoming increasingly difficult for the individual to grasp the whole field of archaeometry with its rapid developments. The aim of the series *Natural Science in Archaeology* is to bridge this information gap at the interface between archaeology and science. The individual volumes cover a broad spectrum of physical, chemical, geological, and biological techniques applied to archaeology as well as to palaeoanthropology with the interested nonspecialist in mind.

The single monographs cover:

- large fields of research
- specific methods of general interest (archaeometric methods of dating, material analysis, environmental reconstruction, geophysical prospecting, remote sensing and data processing)
- materials of interest to the archaeologist, such as sediments, soils, metal and nonmetal artifacts, animal and plant remains and other organic residues
- practical aspects such as sampling and data interpretation
- case studies, to demonstrate the potential and limitations of the various techniques.

Laura Pecchioli
Editor

Archaeoseismology

Methodologies and Case Studies

Springer

Editor
Laura Pecchioli
Institut für Kunstgeschichte, Bauforschung
und Denkmalpflege - Lehrbereich
Technische Universität Wien
Vienna, Austria

ISSN 1613-9712
Natural Science in Archaeology
ISBN 978-3-031-28302-4 ISBN 978-3-031-28303-1 (eBook)
https://doi.org/10.1007/978-3-031-28303-1

This Springer imprint is published by the registered company Springer Nature Switzerland AG
The registered company address is: Gewerbestrasse 11, 6330 Cham, Switzerland

Foreword

> Now Italy itself was afflicted with new disasters, or disasters newly encountered after a lapse of many generations. Cities were swallowed up or smothered—Campania's richly fertile coast; and the City was ravaged by fire, her most ancient shrines devoured, the Capitol itself set on fire by her own citizens—Tacitus.[1]

As the Roman historian Tacitus recognised long ago, disasters have haunted and shaped the Italian landscape for centuries. Earthquakes, fires, storms, invasions, and other calamities have all conspired to modify the Italian environment, both natural and that built. Yet, as much as disasters have seriously impacted societies throughout history, Italy included, the academic study of disasters is relatively recent, only emerging from the 1920s.[2] Understanding the history of disasters and how communities have responded is critical to how we respond to disasters in the future. As a number of historians have recently argued, the past should be considered a type of laboratory where hypotheses which are relevant to the present can be tested in careful ways.[3] The articles which follow all pay tribute to the significant lessons we can learn from the past and collectively focus on how disasters affected the built environment and how communities have responded.

To set the scene, we of course need to start with Rome for not only was it the largest built environment in the Roman Empire but it faced constant and multiple dangers.[4] Here, disasters were not novelties, they were *regularities*. Roman historian Mary Beard has been keen to point out that disasters, particularly floods and fires, 'repeatedly fell upon the place'.[5] So much was this the case that according to one apocryphal tale, a virtuous citizen was given the name Marcus *Caedicius* ('disaster teller') for his earnest and repeated warnings about impending disasters.[6] The spectre of disasters, real and

---

[1] Tacitus, as quoted in R. Jenkyns, *God, Space & City in the Roman Imagination*, Oxford University Press, Oxford, 2013, p. 153.

[2] Samuel Price was one of the early pioneers on disaster research when he conducted research in 1920 on the ship explosion in Halifax, Nova Scotia during World War One which destroyed the town and killed 2000 people. J. Toner (2013) *Roman Disasters*, Polity Press, Cambridge, p. 22.

[3] B. Bavel, D. Curtis, J. Dijkman, M. Hannaford, M. Keyzer, E. Onacker, T. Soens (2020) *Disasters and History: The Vulnerability and Resilience of Past Societies*, Cambridge University Press, Cambridge, p. 1.

[4] Ancient authors give the distinct impression that there was a major fire each night in Rome. G. S. Aldrete, (2004) *Daily Life in the Roman City*, Greenwood Press, London, p. 110.

[5] M. Beard (2015) *SPQR: A History of Ancient Rome*, Profile Books, London, p. 358.

[6] Beard, *SPQR*, p. 121.

imagined, loomed large in the minds of the average Roman and they conceived of them in four main ways: *calamitas* (agricultural disasters and disease); *clades* (military disasters); *casus* (political and military disasters including violent death); and *pestis* (plagues and physical destruction).[7]

How did the Romans respond to disasters? Well, as you likely guessed, it was mixed though there were some common threads. To begin with, emperors generally tried to be pre-emptive but this was mainly in a religious rather than a practical or helpful sense: they consulted soothsayers, made public vows, and offered animal sacrifices to the gods for protection.[8] At a practical level, disaster planning was a primitive and evolving process with solutions typically arising *after* the disaster had occurred, such as the widening of roads after a major fire.[9] All emperors were concerned with what Professor Jerry Toner refers to as 'their architectural legacy' so the destruction of their pet projects was taken very personally.[10] However, their responses to disasters varied and was typically a reflection of their temperament. The Great Fire in Rome of 64 CE gave the emperor Nero the rare opportunity to redevelop large sections of central Rome which he did—but he also saw this as the perfect opportunity to build his infamous Domus Aurea (Golden House). Emperor Titus, on the other hand, after another major fire in Rome, not only eschewed the notion of building a private residence but supplied material from his own residences to restore temples and other public buildings.[11]

Disasters also gave emperors the opportunity to offer patronage which considerably boosted their credibility and power base. Victims needed support, buildings needed to be restored, and social relationships had to be rejuvenated and maintained. With patronage the emperors' responses also varied, some were beneficent to the population at large while others targeted particular segments to win (or at least try) favours. In terms of the people, the wealthy were just as vulnerable to disasters as the plebs. Nevertheless, the community was largely accepting and resilient due to the normality of disasters in their everyday lives.[12]

In relation to restoration an important point is that the ancient Romans were more concerned with 'the antiquity of tradition' rather than the age of a physical building.[13] In Cicero's *Republic*, for example, he compared the Roman state to a masterful painting that had simply faded with age ('iam evanescentem vetustate').[14] He criticises his contemporaries for not renewing the painting in its original, vivid colours. So here, we get the distinct impression that restoration and newness was more important than the authenticity of the building and the so-called charm of deterioration over time. The Younger Pliny agreed with Cicero that mighty deeds and legends of antiquity should be

[7] Toner, *Roman Disasters*, p. 18.

[8] L. Adkins and R. Adkins (2013) *Handbook to Life in Ancient Rome*, Facts on File, New York, 2004, p. 310.

[9] D. Potter, *The Emperors of Rome*, Quercus, London, p. 54.

[10] Toner, *Roman Disasters*, p. 18.

[11] Jenkyns, *God, Space and City*, p. 80.

[12] As Jerry Toner has maintained, 'Risk was, in effect, routinized.' Toner, *Roman Disasters*, p. 20.

[13] Jenkyns, *God, Space and City*, p. 266.

[14] Jenkyns, *God, Space and City*, p. 266.

honoured instead of worrying about the past's appearance in physical form.

Reponses to disasters varied throughout the Roman Empire and was often determined by the scale of the disaster. The catastrophic eruption of Mount Vesuvius in 79 CE is a textbook example. Pompeii was covered in four metres of volcanic ash and Herculaneum was covered in many more metres of volcanic mud, so the idea of restoration was out of the question. As such these towns became forgotten and never rebuilt.[15] However, their rediscovery since the sixteenth century has since offered a multitude of researchers—across an array of disciplines—comprehensive insights into the nature of the disaster, town planning, building techniques, and the quotidian aspects of daily existence.[16] The features of a city do not disappear abruptly, except for rare cases like Pompeii and Herculaneum, though these too have since been reanimated (though not repopulated). Most often, cities fade away slowly and are often in the process of recreating as the following articles compellingly demonstrate across a range of historical periods and settings.

University of New South Wales
Sydney, NSW, Australia

David A. Holmes

---

[15] Potter, *The Emperors of Rome*, p. 64.

[16] B. Reilly (2009) *Disasters and Human History*, Farland & Company, Jefferson, p. 36; Potter, *The Emperors of Rome*, p. 64.

Preface

> I have learnt, excellent Lucilius, that Pompeii has collapsed following an earthquake that has caused damage in the whole surrounding area . . . and has devastated Campania with huge ruins[. . .] A part of the city of Herculaneum has collapsed and even what has remained standing is of uncertain stability; the colony of Nocera, although not seriously affected, has nevertheless something to complain about. Even Naples, which was slightly affected by the great disaster, suffered many losses in its private but not in its public property; some villas collapsed, others, a little everywhere, felt the tremor without injury.
>
> (Seneca, VI—*Naturales Quaestiones*)[17]

Throughout history, earthquakes in the Mediterranean have caused the destruction and redevelopment of the built environment. There are numerous references to catastrophes in the surviving documents of the ancient Greeks and Romans.[18] Relative seismic effects have inspired various theories regarding the earthquake-induced collapse of entire civilisations. There have been many attempts to try and define a framework, both geographically and historically, for the destruction of layers, the abandonment of buildings, and cultural changes.[19] Due to the scarcity of sources, archaeological and geological data are often used to supplement historical evidence.

A current topic much debated in various research fields is urban resilience. If we reflect on the history of how governments and societies have responded to disasters, we see that our accumulative knowledge has improved, but still, often, the vulnerability and the risk are underestimated.

If we look at the history of city development and the rise of civilisations through time, it would be evident that cities have been developed based on people's needs, constantly evolving and adapting to internal and external conditions. Analysing the vulnerability and resilience of past societies can show how we can use history to understand better how to deal with

[17] *Naturales Quaestiones* was written between 62–64 CE in eight books. The dating is clear from references to the Campanian earthquake (62 CE) and mention of the comet of 60 CE that had caused some consternation among Nero's supporters at Rome. While the manuscripts preserve the text in three different orders.

Book 6: Earthquakes—Seneca begins by reviewing the recent Campanian earthquake before investigating possible elemental causes of earthquakes. The conclusion focuses on diminishing one's fear and providing remedies for the traumatic aftermath of these natural occurrences.

[18] Joseph D. Hughes (2012) Responses to Natural Disasters in the Greek and Roman World. Forces of Nature and Cultural Responses. Nov 10:111–37. DOI: 10.1007/978-94-007-5000-5_7. PMCID: PMC7121757.

[19] Stathis Stiros (2016) *On the historical role of earthquakes in Antiquity* In: *Géoarchéologie des îles de la Méditerranée* [online]. Paris: CNRS Éditions (generated 28 September 2022); http://books.openedition.org/editionscnrs/28623

catastrophes. The expectation that central governments played a role in disaster recovery, as in the case of earthquakes, can be traced back to the actions of Roman emperors. *David A. Holmes* provides a historical introduction to this collection, teasing out several themes dealt with by the authors, such as societal responses to earthquakes. A historian's perspective underlines the relevance of studying past disasters and how it contributes to a better understanding of how societies functioned. We cannot properly understand past disasters without providing a historical context and examining historical documents.

Archaeoseismic research provides data and information on past earthquakes but is limited by the lack of ongoing discussions about methodology. In the interdisciplinary approaches, when examining catastrophic events, methodological issues play a fundamental role.

Among relevant aspects, *Cairoli Fulvio Giuliani* focused on 'critical surveying' that should demonstrate the inter-relationships between the whole and individual components of the architectural object. An invitation to review the way of reading monuments, without delegating to modern technology, can be based on a critical methodology and approaches. That also often requires the development of new research methods common to multiple search fields.

In this context, archaeoseismological studies can reveal the possible natural causes of a site's destruction and contribute to the broader field of environmental archaeology, which seeks to define the history of the relationship between humans and the environment. Finally, these studies can add to regional seismic hazard assessment by improving information to date and integrating the implemented catalogues. Archaeologists, geologists, geophysicists, historians, seismologists, engineers, and architects from different countries demonstrate impressively that recording and interpreting ancient natural disasters cannot be done by scientists from a single discipline but through an interdisciplinary partnership.[20]

The collection provides an overview of research into archaeoseismology, making new contributions through innovative ideas on various topics. The publication can be an illustrative introduction to better understand the complexity of interpreting seismic effects on ancient and modern masonries, particularly for students with an open-minded.

Some authors represent the discipline's most influential founders and pioneers. Other colleagues have developed methodologies and conducted projects on an urban and building scale through accurate models of interpretation of geological, historical, seismic, and archaeological data. The idea for this publication was born after an international workshop in Berlin in 2018,

[20] On the occasion of the European Seismological Commission 36th General Assembly in Malta, in collaboration with two colleagues seismologists, a new working session on cultural heritage and earthquakes was accepted. Several representatives of archaeoseismology also attended the meeting. On this occasion, it was also possible to propose a special issue to the Journal of Seismology as guest editor, Cultural Heritage, and Earthquakes: bridging the gap between Geophysics, Archaeoseismology, and Engineering: https://www.researchgate.net/publication/342279463_Cultural_heritage_and_earthquakes_bridging_the_gap_between_geophysics_archaeoseismology_and_engineering

supported by the Gerda Henkel Foundation.[21] Concerning the event, a summary of Stathis C. Stiros provides a guide on the relevant issues in the discipline:

On the Role of Ancient Earthquakes in the Building Style and the occupation patterns

Have earthquakes influenced building style and occupation patterns in various regions and periods? Despite the recent progress in studying various aspects of ancient earthquakes, a clear answer to this question is not easy. In several cases, archaeological and historical data provide evidence of complete re-organisation of town planning, changes in the building style, and even shifting of inhabitation centres after ancient and sub-modern earthquakes. In other cases, on the contrary, urban patterns seem insensitive to earthquakes, but occasionally, with a reversal of the long-term pattern.

There is also evidence that the persistence of old styles/patterns or the selection of new ones is controlled by a 'cost-benefit analysis' of various factors (security, natural resources, broader politico-economic context, etc.) by catalysing the role of earthquakes facilitating impending changes, by migration, and by episodic environmental effects (especially loss of fresh water supply). Understanding these processes requires a detailed and reliable database from various periods and regions and synergy between different disciplines, but also refinement and expansion of results of Historical Seismology and Archaeoseismology, which are not without problems for the moment. Just notice that the theory of aseismic oases inside broad zones of deformation and seismicity collapsed only after a 1995 M6.5 Greek earthquake. In critical regions, the apparent earthquake rates with time seem inconsistent.

The main reason for these problems is imperfections in the analysis of literary sources. That apparent, highly catastrophic event is usually uncritically adopted, ignoring the limitations of ancient authors and the ideological and political perception and exploitation of physical events through different generations, leading to missed or biased, even physically impossible events. On the other hand, studies of earthquakes based on archaeological evidence tend to a priori adopt deterministic approaches. In contrast, the probabilistic and conditional character of palaeoseismological studies is usually ignored, along with each case study's engineering and socio-political context. Because of these approaches, small deformations of buildings are unconditionally regarded as seismic effects. In contrast, Kourion-type destruction layers are regarded as evidence of deliberate actions, and dubious correlations of damages are attempted.

Uncertainty and bias in palaeoseismic data are inevitably reflected and amplified in the study of the cultural and historical impacts of earthquakes and their impacts on the built environment.

Stathis C. Stiros
Department of Civil Engineering, Patras University, Patras 26500, Greece

Historical sources preserved in an archive remain an essential basis for the seismic reconstruction of a place, especially when one does not know the exact location. The contributions in Città di Castello (*Giovanetti*) and the study of historical seismicity in Vienna (*Bruno* et al.) show how essential a detailed and cross-referenced analysis of sources is. This is especially the case for an accurate interpretation and for adopting a constructive methodology repeated over time.

The case studies of the geographical areas presented reach as far as Greece, which has been significant for catastrophes and seismic events since antiquity. The contribution to Roman Corinth and the wider region (Greece), written by *Stiros*, is a significant reference in archaeoseismology. The Roman archaeological stratigraphy in this area is traditionally explained in terms of

[21] International Workshop, 15–16 November 2018 in Berlin: *Historical Earthquakes dialogue between Archaeology, Geology, Archaeoseismology, and Building Research*: HTTP://HIST-QK.NET/INDEX_EN.HTML

earthquake effects. Still, they do not represent events derived from the study of ancient remains, events using specific criteria.

Several basic research methods are illustrated: the stratigraphic reading of walls and analysis of collapsed mediaeval buildings, the identification of repairs after earthquakes (*Arrighetti* et al.; *Montabert*), and in particular, the integration of methods with digital archive technologies (ANR RECAP) applied in the archaeological context of Pompeii (*Dessales* et al.). Technological solutions recall ancient construction techniques and reinforce parts of the building that are resilient to disasters. On the other hand, the knowledge and interpretation of seismic effects pose difficulties: a comprehensive and exhaustive classification applied currently in different archaeological cases has been undertaken by *Pascua* et al. Their contribution is based on a methodology that emphasises the documentation and surveying on-site for identifying the main parameters relating to an earthquake.

How important is preserving collective experience in dealing with a local earthquake culture? In order to increase and improve our knowledge of past seismicity, it is essential to consider archaeoseismological studies to characterise the seismogenic behaviour of a region. There is a direct connection between intuition and engineering based on the science of construction, which must be recovered. Anti-seismic techniques inherent to the local seismic culture are described in detail in *Pecchioli–Cangi* evaluated. Numerous case studies on preventive solutions linked to the Roman art of buildings have been presented in several articles and emphasise what the main tools should be, including direct on-site analysis, archival work, excavation reports, stratigraphy, and masonry mechanics. Today, with the new conservation codes and decrees, there is an attempt to integrate local solutions to the normal practice of recovery and consolidation, but the unpreparedness to decipher the seismic traces and the 'not-knowledge' of the mechanical characteristics of the materials still prevails.

In localised reinforcement in structures, foundations are often the structural system responsible for the response of a building to a seismic impact. According to *Diosono* et al., one of the latest discoveries relative to the temple foundations and podiums in Greek is illustrated as a result of interdisciplinary studies on the seismic shield. Also, in ancient Rome, wooden roof structure solutions through a non-linear 2D mechanical model for FEM with *Fei* et al. are analysed in depth to evaluate the mechanical behaviour.

Keeping within the Roman framework, the contribution by *Dörings-Williams* and *Albrecht* shows a fascinating construction analysis to reconstruct a seismic chronology of a relevant Roman monument. Theories focus on seismic evidence on-site relative to the collapses in Late Antiquity as an assessment of the possible causes of destruction.

Berlin, Germany Laura Pecchioli

Acknowledgments

Many thanks to Annette Buettner and Bibhuti Sharma for the chance to publish this collection with *Springer Nature*.

I would also like to thank the authors for their rigorous research and for sharing their innovative ideas and results. Their contributions give value and relevance to the volume.

Thanks to my colleague David A. Holmes who edited these papers and provided a splendid historical introduction.

A special thanks to Cairoli Fulvio Giuliani, for mentioning Antonino Giuffrè, who emphasised the essential need to recognise and acquire that knowledge of structural mechanics in analysing historical architecture. Giuliani highlights how today we underestimate the importance of the real datum. One can observe sadly how the profound diversity is still often unresolved between digitising the object and understanding its structural mechanics. An unfortunate confusion too widespread in archaeological studies of the means, and which too often tends towards the narrative.

My heartfelt thanks to those who have developed me as a researcher and supported my teaching in archaeoseismology: Stathis Stiros, Fabrizio Galadini, Klaus G.- Hinzen, Emanuela Guidoboni, M. A. Rodriguez Pascua, Bruno Helly and Donat Fäh. Synergies that still cooperate in a fruitful way today.

Introduction

While my academic training only partially qualifies me for presenting a collection of texts such as this one, my age and experience in the masonry field oblige me not to shirk.

Antonino Giuffrè wrote in 1988: 'It is evident that ancient masonry buildings cannot be seen as a whole structure, but as a set of structural parts, each of which presents an autonomous resistant mechanism that must if need be, be self-verified'.

The whole difficulty of the subject matter is expressed in a few lines.

Traditional archaeology, involved between a necessarily epidermic and subjective description of residual buildings and hard-to-acquire technical knowledge, has largely overlooked and neglected the evidence of the subject's complexity.

The underestimation of the importance of the real datum, the profound diversity, still often unresolved, between digitising the object and understanding its structural mechanics, and the unfortunate confusion still too widespread in archaeological studies of the means (survey) with the end (knowledge) all too often tend towards the narrative.

This allows for the strange phenomena of practical fallout.

I will cite just one example that may seem irrelevant, but which is threatening both the aesthetics and, in perspective, the structure of many of our historical centres.

I am referring to the senseless interventions that, in the absolute indifference to plant engineering (out of optimism, incompetence, crude speculation, or simple ignorance?) in the widespread ancient and old buttresses reduce, when not completely nullify, the 'out-of-plumb' response of the façades by opening up housings for various types of 'gas/water meters'. I have counted up to six of these boxes, contiguous ones capable of nullifying the function of the buttress altogether.

It is well known that spurs, almost always a remedy for seismic out-of-plumb, have proved efficient in blocking the out-of-plumb itself and subsequent seismic episodes. Anyone can imagine the possible consequences of their abolition.

In addition to the aesthetic damage, nullifying their structural function does not induce optimism.

Knowing architecture, like ancient topography and town planning, requires specific attention in addition to critical judgment, like any other field.

In this case, to the course of load flows, the changes in the equilibrium of masses over time, the study of cracks, and structural mechanics as a whole.

Therefore, any form of reading of the artefact, whether structural, building, morphological, metric, technological, etc., needs 'surveying'. But that it be critical, not just a simple technological operation, to be exhibited as a 'scientific' garment to accompany an article.

Substantially useless is instead the 'description' of the object, which exposes the scholar to the temptation of having found what he was looking for (see the Wall of Romulus).

This is why the survey must reproduce and, above all, demonstrate the inter-relationships between the whole and the individual components of the architectural object: first, the how and why of the object, and only then the when and who realised it and the congruity with the correct interpretation of the literary sources.

Given that a critical survey is an operation that cannot be fully objective, one will understand the difficulty of proceeding, especially when the actors (the technician and the scholar) are distinct entities.

This perhaps partly justifies the annoyance I have always felt at how archaeology generally treats architectural structures, which constitute the most important historical archive for millennia of building culture.

The critical cultural crisis that is beginning to manifest itself in work such as this directly concerns historical building, the focus on structural mechanics, which has nothing to do with the widespread surface masonry stratigraphic units (UUSSMM), the epidermic and useless mapping of the brickwork, the disquisitions on 'reticulated' and 'quasi-reticulated', etc.

For a very long time, damage and its repair, as well as changes in use, sometimes caused radical changes in the original load flows, which could even date back to the original design (cf. Pliny the Younger, Carteggio con Traiano, 2, 17; 39–40), until the building was reduced to ruins.

One can thus understand how the pretense of understanding the multiple, often contradictory facets of the history of a two-thousand-year-old building through the study of surfaces instead of skeletons leads nowhere.

To anaesthetise that still-dynamic structure into a serialisation of cards based on observations limited to the epidermis (UUSSMM), to neglect the value and significance of the crack paintings that constitute the interpretable archive of the vicissitudes of the masonry framework and its changes over time, is to renounce consulting the fundamental historical document of archaeological architecture, in other words, to prefer opinionism to knowledge, subjective assertion to a demonstration.

As long as the problematic paradigm of attuning the scholar with the original designer in the structural sphere is not applied in the study of ancient topography, architecture, and town planning, these disciplines, as is the established tradition, will tend more and more towards the subjectivity of description in the broader framework of an opinionism that, even if it is sometimes supported correctly by philological suggestions, evaporates more often in the scant consideration of fact and appears to be built on fictional, albeit literary, foundations.

In front of an ancient wall, one always finds two very distant characters: the master mason who built it with an experience stratified over millennia, a repository of a rich specific cultural heritage, and the archaeologist who, at an enormous distance in time, tries to understand something about it.

The cultural characteristics of the original designer, although fundamental, no, those, if anything, come later as a philological consequence.

On the road to a new paradigm, the archaeologist must equip himself since his opinion, preconception, or irresistible desire to find what he is looking for are not enough.

Nor is it enough to accompany one's work with others, especially if it is technologically excellent. We are all aware that we know how to find what we are looking for. The difficulty is demonstrating that we have done so with concrete data, critically interpreted and not simply protected by a technology that, if not actively participated in, remains only an alibi.

An important fact emerges from this accessible collection, the emergence of a new sensitivity for historic buildings, which requires the contribution of disciplines from different backgrounds united by an interest that can be reduced to two fundamental concepts:

- First, the recognition of an extensive historical archive of built remains for the knowledge of ancient architecture, and also through the fundamental and complex research field of seismology.
- Second, the indispensable confluence of differentiated specialisation: wherein technicians specialised in the mechanics of structures and seismologists are interested in an adequate course correction in the research of the complex experimental history of construction developed on projects, error corrections, and repairs from the distant origins up to at least the mid-nineteenth century. Evidence that I was fortunate enough to see still survives in some reconstructions after the Second World War and, perhaps, in some isolated cases, still lives on.

Moreover, the only way to attempt to understand the work of 'that' master builder I mentioned is by surveying, possibly in the first person, I repeat, understood as a critical reading, a continuous dialogue with the wall, and carried out by someone who studies that wall, because when the archaeologist resorts to 'technician' he has two possible ways: either he ends up interpreting the technical product rather than the structure, or the technician undergoes surveying what the scholar is convinced in good faith that he has understood.

The only certainty is that the surveyor accumulates more information than ends up in the survey, and he will understand the problem best, regardless of what the client thinks.

Not only that, but the technological extremists also help to delude those unfamiliar with surveying that it is a simple geometry job.

Knowledge, in reality, is approached through constant and direct dialogue with the individual parts of the building, recognising their interrelation in the context of a project as far apart in terms of expertise as in this collection.

Università La Sapienza, Rome, Italy Cairoli Fulvio Giuliani

Contents

About the Editor

Laura Pecchioli PhD (European); Gerda Henkel Stiftung: Research Project (2017–2021): *Präventiver Erdbebenschutz und historische Reparaturen von Mauerwerk. Drei Fallstudien: Piccolo Mercato, Capitolium und Insula del Serapide e degli Aurighi* (*Ostia*): https://lisa.gerda-henkel-stiftung.de/der_puls_der_erde?nav_id=7556 (habilitation theme and forthcoming publication); Research Fellowship-Università degli Studi ROMA TRE, Dipartimento di Architettura e Restauro—Studi Urbani di Roma (2022–2023); Humboldt Universität, Institut für Archäologie, Klassische Archäologie/Winckelmann-Institut; Lehrbereich 2019-2020: 1. Einführung in die Archäoseismologie (SoSe-BA); 2. Historische Erdbeben und ihre Auswirkungen auf die Entwicklung von Bautechniken (WiSe-MA); Ostia Forum Project (OFP)—http://ostiaforumproject.com/;Technische Universität Wien, Institut für Kunstgeschichte, Bauforschung und Denkmalpflege; Lehrbereich 2021-: Architekturdokumentation und Präsentation/Transdiziplinäre Interaktionen, Einführung in die Archäoseismologie und Archäometrie (WiSe-MA) https://tiss.tuwien.ac.at/course/courseDetails.xhtml?dswid=6621&dsrid=40&courseNr=251730&semester=2021W; (Lecturing in Archaeoseismology at the TUVienna since 2021). ICOMOS-ICORP International Scientific Committee on Risk Preparedness; ISCARSAH, International Scientific Committee on the Analysis and Restoration of Structures of Architectural Heritage; WATCH, World Association for the Protection of Tangible and Intangible Cultural Heritage during Times of Armed conflicts; ESC-Working group, European Union Seismological Commission.

Part I

Stratigraphic Analysis and Seismic Evidence

1 Stone Materials, Processes, and Mortars in Archaeoseismology Studies: The Case of Mugello (Tuscany-Italy)

Andrea Arrighetti and Fabio Fratini

Abstract

Mugello is an intermontane basin with medium-high seismic risk located in the Apennine Mountains on the border between Tuscany and Emilia Romagna. The territory is characterized by numerous long-lasting settlements which comprise well-preserved historical buildings. Between 2010 and 2014, an area of Mugello was involved in the project "Archaeology of architecture and seismic risk in Mugello", research focused on evaluating the archaeological analysis process as a form of prevention and protection of mediaeval buildings from seismic risk. This research describes the central role played by the study of the provenance, processing, and installation of the stone materials used for the construction and modification of architectural structures during the period twelfth to the twentieth century. It also illustrates the potential that the study of bedding mortars can have in archaeoseismology. These materials contain much technical information, starting from the raw materials used for their realization (stone for the production of lime, aggregate) to arrive at the mix design (binder/aggregate ratio, quantity of water). Therefore, their study makes it possible to contribute to distinguish and characterize masonries even when they are built using the same type of stone material and with similar processing and finishing.

Keywords

Archaeoseismology · Architecture · Building techniques · Chronotypology · Mortars

1.1 Introduction

Mugello is a medium-high seismic risk area located in the Apennine mountain range on the border between Tuscany and Emilia Romagna. The territory has been continuously inhabited since the Middle Ages and is characterized by well-preserved historic buildings, most of which with a religious function.

An area of Mugello between the municipalities of Scarperia, Borgo San Lorenzo, Vicchio, and Barberino was examined between November 2010 and March 2014 for the project "Archaeology of architecture and seismic risk in Mugello". The project provided an archaeological analysis of mediaeval buildings in relation to their vulnerability to seismic risk (Arrighetti 2015). Among the results that emerged from archaeological investigations, considerations concerning the supply and use of building materials for the construction and modification of architectural structures

A. Arrighetti (✉)
École Normale Supérieure, Université PSL, Paris, France
e-mail: andrea.arrighetti@ens.psl.eu

F. Fratini
CNR ISPC (Consiglio Nazionale delle Ricerche, Istituto di Scienze del Patrimonio Culturale), Florence, Italy

L. Pecchioli (ed.), *Archaeoseismology*, Natural Science in Archaeology,
https://doi.org/10.1007/978-3-031-28303-1_1

played a central role during the period twelfth to the twentieth century.

One of the most interesting aspects regarding the construction history of the buildings was the potential to identify, catalogue and in some cases to date the structural systems that characterize them. The sources used in the research were essentially of two types: the first, historical sources, useful for providing the historical context and timeline for specific construction events; second, the stratigraphic reading of the exposed masonries, which provides insight on the construction phases of the analysed artefact. Therefore, this type of analysis made it possible to propose a characterization and dating of the anthropic operations of modification of the building due to natural events, such as earthquakes (Arrighetti 2022).

The chrono-typologies of the proposed construction systems made it possible to catalogue and date the different masonries. Once compared, these masonries offered the possibility of hypothesizing the use of specific expedients in their construction. At the same time, the definition of a chrono-typology of restoration interventions in the seismic area led to hypothesize the role of workers and customers in the post-earthquake phases. This practice was considered to be essential to hypothesize the presence or absence of a Local Seismic Culture (Pierotti and Ulivieri 2001; Pierotti 2003) in the know-how of the builders in the area investigated.

1.2 Methodological Lines and Analysis of the Artefacts

The project "Archaeology of Architecture and seismic risk in Mugello" was characterized by a methodology of context analysis based on the archaeological approach to the study of mediaeval architectural artefacts (Brogiolo and Cagnana 2012). A first systematic survey of the study area (Fig. 1.1) led to the identification of numerous buildings to be analysed. Then a total of 12 architectural complexes representative of the quality of the exposed masonry structures (capable of allowing a stratigraphic analysis of the construction phases) and representative of the period of interest (from the twelfth to the seventeenth century) were selected. The deepening in the analysis of buildings was set up by groups of activities,

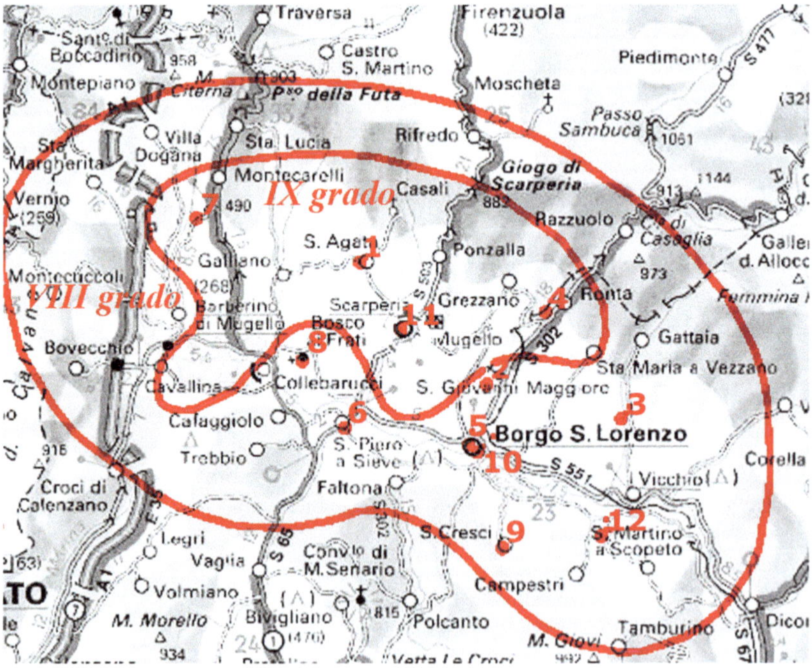

Fig. 1.1 The 12 buildings analysed and the boundary of the study context, the latter corresponding to the VIII degree of Macroseismic Intensity hypothesized by the National Institute of Geophysics and Volcanology for the 13th earthquake of June 1542 (Ferrari and Molin 1986)

Fig. 1.2 The image shows an example of the stratigraphic reading carried out by Wall Stratigraphic Units (on the left) and groups of construction periods (on the right). In particular, the figure presents the stratigraphic analysis of the east elevation of the church of Sant'Agata del Mugello (FI)

cuminating at an analysis of Stratigraphic Units only when stratigraphic complexity required a very detailed study (Fig. 1.2).

After the stratigraphic analysis a relative chronology of the construction phases was realized, which led to the first interpretation of the archaeological data and a chrono-typological characterization of the individual masonry construction techniques. Then a review of both published and unpublished written sources in local and national archives and libraries was conducted, making it possible to propose absolute chronologies for some phases. The final step was interpreting the processed data in relation to the 12 artefacts in the specific regional setting. Thus, it was possible to understand better the changes in the settlement dynamics of the area and the related social, economic, and political aspects.

1.3 The Supply and Use of Building Materials from the Late Middle Ages and Early Modern Age

The Mugello is morphologically defined by a depressed central area, crossed in its southern part by the Sieve River. This main area is made up of fluvial lacustrine terraces. To the north is the Apennine chain, consisting mainly of sandstone geological formations; to the south is the Monte Giovi—Monte Senario chain, consisting of sandstone formations in the eastern area and limestone marly formations (Pietra Alberese) in the western area (Fig. 1.3).

These geological formations supplied the construction materials, represented by the Alberese stone (marly limestone) (Fratini et al. 2020), the Pietra Serena (sandstone) (Fratini et al. 2014), the Monte Senario sandstone (Cipriani et al. 2005), and clay for the bricks. To these materials, we should add the Green Marble of Monte Calvi (serpentinite, Fig. 1.4) (Fratini et al. 2022) and the over-burned and vitrified bricks (Fig. 1.5) used sporadically as a decorative material.

Finally, the presence of limestones and sandstones of the Sillano Formation is used as construction materials, always together with the Alberese stone. By comparing the chrono-typological data, it is possible to identify four historical periods linked to construction periods of significant interest. Each phase seems to be characterized by choice of one or more materials depending on factors of different nature: economic, political, or simply practical. In particular,

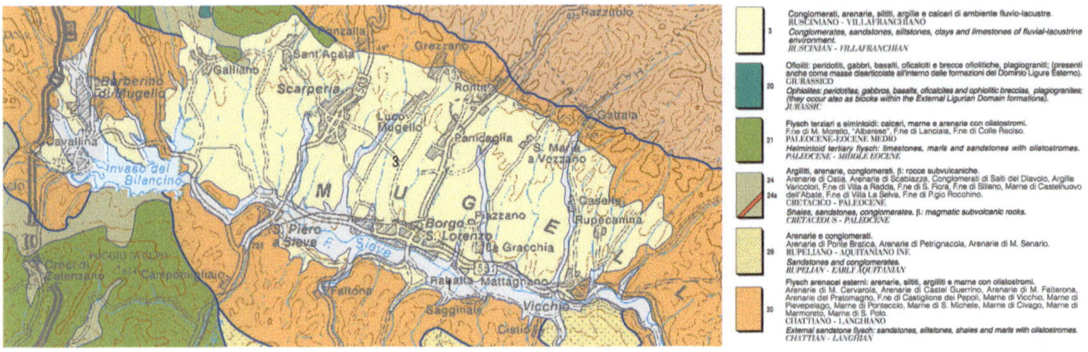

Fig. 1.3 Detail relating to the Mugello of the Geological Map of Tuscany (geological map 1:250,000 modified after Carmignani and Lazzarotto 2004)

the use of construction materials falls within the following periods:

First period (tenth–twelfth century): the choice of the stone materials seems to depend strictly on the lithologies present in the vicinity of the building rather than on the evaluation, by workers and clients, of the characteristics (constructive and decorative) of the various raw materials (Fig. 1.6). In this period, the buildings of almost all 12 religious structures involved in the research can be included. A clear demarcation of use is evident between Western Mugello with Pietra Alberese and Eastern Mugello with Pietra Serena and Monte Senario sandstone.

Second period (thirteenth–fourteenth century): starting from the first half of the thirteenth century, a new construction material began to be used: bricks from a particular area of the Mugello region. The bricks were for a religious client that probably employed specialized workers. The hypothesis of a possible interaction between the introduction of

Fig. 1.4 Chessboard composed of sketches of Green Marble from Monte Calvi and Pietra Alberese present on the north elevation of the Pieve di Sant'Agata del Mugello

Fig. 1.5 Vitrified bricks present as a decorative element in the joints of the bell tower of the Pieve di San Giovanni Maggiore

the new construction material in Mugello and the presence of non-local workers seems confirmed by the stylistic and archaeological analysis of the bell tower of the Pieve di Borgo San Lorenzo, whose architectural typology, as already hypothesized by other scholars (Moretti and Stopani 1974), could suggest influences coming from the Lombard region (Figs. 1.10 and 1.11).

In the rest of Mugello, there is a continuity in the use of Pietra Serena and Pietra Alberese (Fig. 1.7). The phenomenon is probably linked to the constant exploitation of the same supply areas for the raw materials, perhaps related to the presence of the same workers and clients of the previous period.

Third period (fifteenth century): this phase is a transition between two great construction moments affecting the Mugello area, namely the extensive restorations carried out in between the thirteenth and fourteenth centuries and the post-seismic reconstructions from

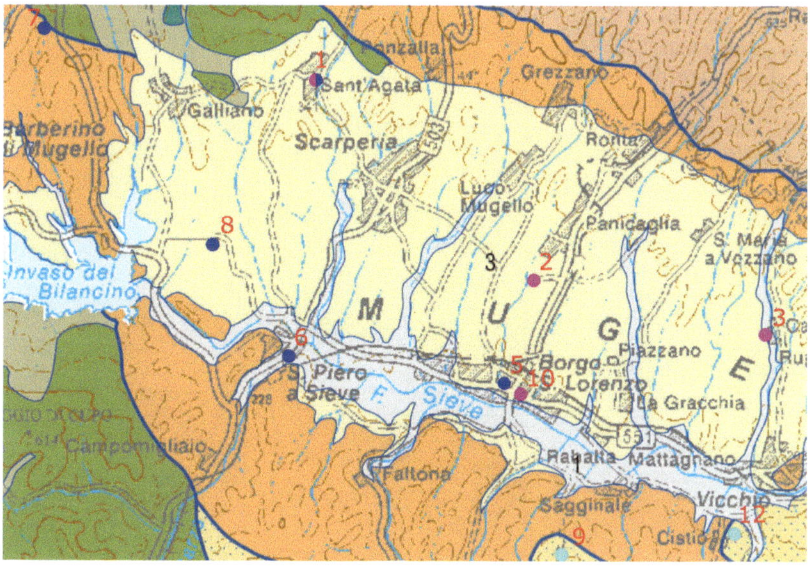

Fig. 1.6 Building materials used in Mugello from the tenth to the twelfth century: Pietra Alberese (blue colour), Pietra Serena (magenta colour), and Monte Senario sandstone (cyan colour) (geological map 1:250,000 modified after Carmignani and Lazzarotto 2004)

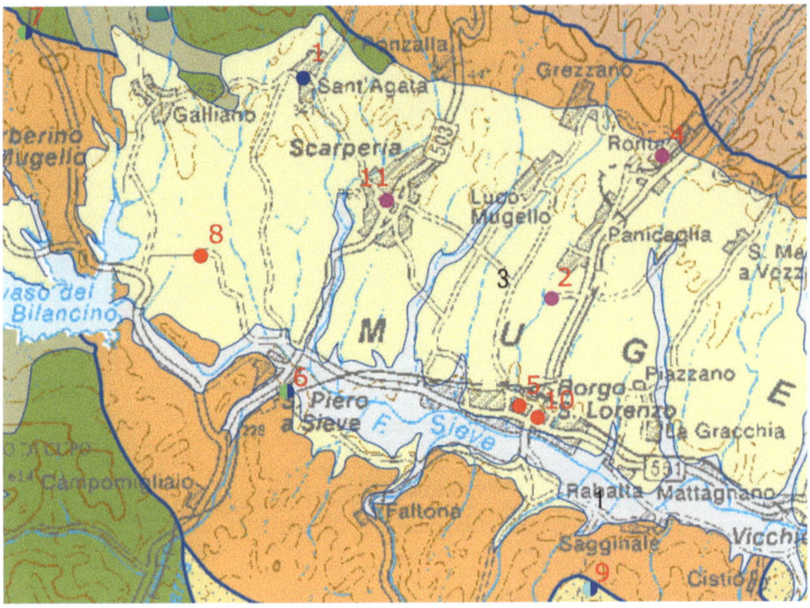

Fig. 1.7 From the thirteenth to the fourteenth century: Pietra Alberese (blue colour), Pietra Serena (magenta colour), Brick (red colour), and Monte Senario sandstone (cyan colour) (geological map 1: 250,000 modified after Carmignani and Lazzarotto 2004)

1542 to the first half of the seventeenth century. There is a continuity in the use of stone materials in the areas described in the first period, flanked by an introduction of bricks for some restorations carried out in the Palazzo dei Vicari in Scarperia (Fig. 1.8). In the latter case, a link between a very wealthy client, the Medici family, and the specialized workers involved in the construction of bricks during the second period can be hypothesized.

Fourth period (sixteenth–early seventeenth century): the restorations and reconstructions carried out following the earthquakes of 1542, 1597, and 1611 date back to this period. It is interesting to note the abandonment of the Alberese stone, in contrast to the extensive

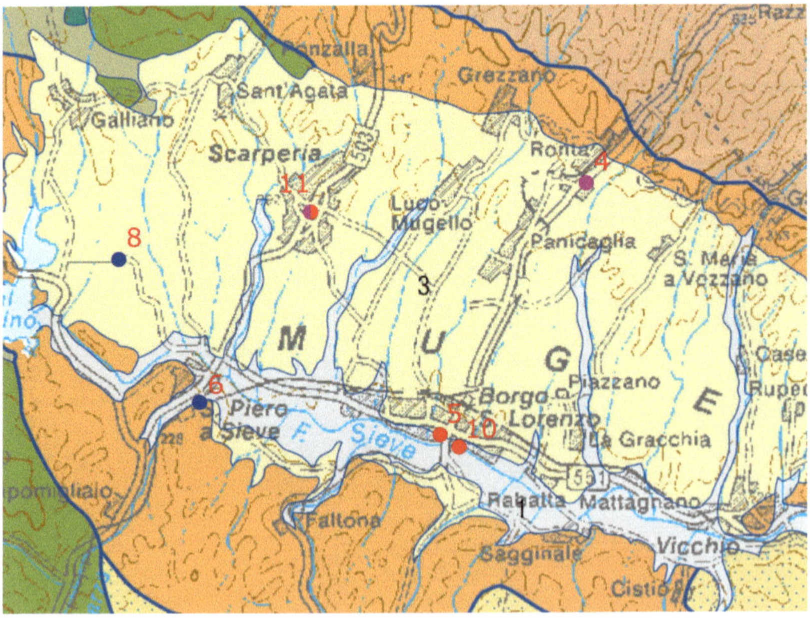

Fig. 1.8 In the fifteenth century: Pietra Alberese (blue colour), Pietra Serena (magenta colour), and brick (red colour) (geological map 1:250,000 modified after Carmignani and Lazzarotto 2004)

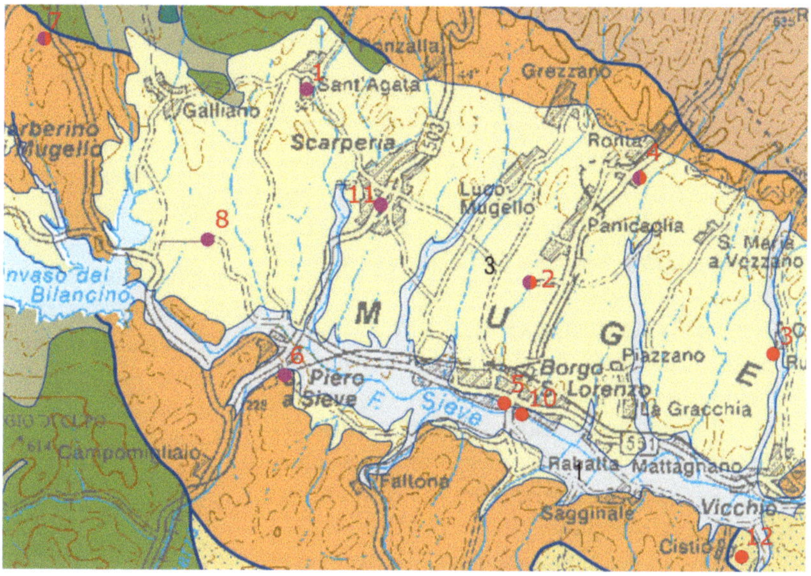

Fig. 1.9 In the sixteenth and early seventeenth centuries: Pietra Serena (magenta colour) and brick (red colour) (geological map 1:250,000 modified after Carmignani and Lazzarotto 2004)

use of the Pietra Serena and bricks (Fig. 1.9). The reason for this preference is unclear. A possible explanation could be that from the fifteenth century onwards in Florence and the surrounding area, the Pietra Serena was brought into vogue with the new Renaissance style, becoming the stone material predominantly used as construction material and for the production of decorative or architectural elements. Likewise for bricks, which were widely adopted as a construction material in buildings in Florence and its countryside from the 16th century onwards.

In summary, the historic buildings of Mugello, in the first period of construction of religious structures (tenth to twelfth centuries) and the constructions or restorations carried out in subsequent periods strictly depend on the availability of local raw materials. Unfortunately, a study on the historical quarries present in the area is still lacking. The bricks represent a characterizing element that allows connecting the use of a specific raw material with the needs of a client. The hiring of specialized workers probably arises from the significant change in the constructive culture of Mugello, not only for introducing a new building material in architecture but also for the innovations in its production.

1.4 A Chrono-Typological Atlas of the Traditional Building Techniques of Mugello

By comparing the data obtained from the characterization of the masonries identified in the analysis of the sites, it was possible to formulate some considerations regarding the *Wall Construction Techniques* in the various historical periods, thus allowing the drafting of a "Chrono-Typological Atlas". The first phase of the work involved cataloguing the different techniques, performed by identifying, characterizing, and dating the different masonries in relative chronology by the stratigraphic reading of the architectural artefacts. This operation was followed by comparing the data emerging from the archaeological analyses and the dating proposed from historical sources. Through this process, it was possible to offer an absolute chronology of five wall techniques. The criteria adopted in the classification of these masonry construction techniques were derived from Fiorani (1996a); Fiorani (1996b) for the processing of materials; Parenti (1987) and Mannoni (2005) for the classification of masonry texture; Bessac (1986) for the finishing of the stone elements.

From the examination of the different Wall Construction Techniques, it is possible to propose the following remarks:

Masonry texture: this seems to be one of the distinctive parameters for understanding the constructive characteristics adopted in Mugello in the different historical periods. *Regular laying* is attested in eight cases, in a chronological span ranging from the eleventh to the early nineteenth century. The laying in *puddled courses*, regular ashlars made of Pietra Alberese and sandstone (Pietra Serena and Monte Senario sandstone), and pebbles of Pietra Serena are attested in four cases, and their use is concentrated from the eleventh to the thirteenth centuries.
The masonry texture worsened in the twentieth century, with irregular laying attested in most cases. Only the monumental buildings located in the city centres (e.g. Pieve di Borgo San Lorenzo) seem to deviate from this phenomenon.
The joints and laying beds seem to be characterized by relatively large dimensional ranges in the eleventh century (from 2.2 to 0.5 cm), arriving then at much smaller dimensions in the twelfth century (from 1.2 to 0.2 cm) and finally to a sort of standardization for constructions of 1420 and 1525–26 (from 0.7 to 0.3 cm). From the eighteenth century onwards, a negative trend is observed that leads to an increase in thickness of the joints, up to total dimensional irregularity for the twentieth century. It is also possible to note that Pietra Alberese, when used as the only construction material, is characterized by a joint thickness between 1 and 0.3 cm. In contrast, when used with the limestones and sandstones of the Sillano Formation, it shows joints with dimensions of up to 2.4 cm.

Processing: regarding the processing of the materials concerning the Pietra Alberese, the most representative material among those of the studied buildings, no worsening or improvements are noted in the period examined. There is, in fact, a substantial continuity in the shaping of the ashlars. The heights and widths are rather irregular, except in two cases characterized by extremely standardized sizes. The use of unprocessed Pietra Serena pebbles has been observed continuously from the eleventh century to the restorations carried out in the twentieth century. Finally, in the twentieth century, there was an absence of processing of the materials used in the analysed buildings.

Finishing: the surface finishing of the external facing was mainly reconstructed by observing the walls realized in Pietra Alberese, where, thanks to the absence of surface decay, traces of the tools used in the finishing phase remained. The most used tools seem to be those with a flat blade, with continuous use from the twelfth to the nineteenth century. The external flattening of the surfaces is sporadic, observed only in the twelfth and thirteenth centuries walls of the Parish Church of Sant'Agata. The use of the subbia or other pointed instrument is observed only in two cases, one in the twelfth century and the other in the fifteenth century.

Bricks: the dating of the use of bricks in Mugello is unreliable. The measurements were obtained from the analysis of the external and internal masonries of only three Architectural Complexes on Construction Phases dated in absolute or relative chronology.

These data are extremely heterogeneous and let us suppose the presence of non-standardized brick production methods in Mugello. The wide range in the dimensions excludes the possibility of obtaining types of reference from the analysed *Wall Construction Techniques.* For this reason, the data obtained from the brick facings were not used to interpret the evolution of the construction systems in the studied context. On the other hand, the information processed supported the interpretation and dating of the masonry construction techniques in lithic materials, consisting of much more solid databases.

By relating the characteristics of the different analysed techniques, it is possible to propose a "Chrono Typological Atlas of the Mugello Building Techniques", a tool capable of highlighting the manufacturing and laying processes used by the masons in the various historical periods. In this way, it is possible to make assumptions about the constructive characteristics of the buildings in Mugello and the economic-political factors at the base of the choices made by the clients.

1.5 The Role of Local and Specialized Workers in the Construction of Buildings

To reconstruct the role of the active subjects in the construction and interventions carried out on the buildings of Mugello in the various historical periods, it is important to integrate the data obtained from the analyses carried out on construction techniques, supply and utilisation of building materials, with the results emerging from the historical-archaeological analyses on the historical contexts of reference.

The results are divided into chronological periods which concern the main construction periods of the area:

Eleventh–twelfth century: for this period, the data on the supplying and use of the main construction materials, the latter mostly derived from raw materials quarried on site, do not seem a criterion that can be considered in order to understand the choices by the workers and clients. Instead, a distinctive element is the adoption, albeit sporadic, of materials for specific, decorative purpose available nearby from the site of construction. This is the case of the Monte Calvi Green Marble for the Pieve di Sant'Agata, an element that denotes a very specific will on the part of the clients in choosing the decorative material to be used. The same is for the over-burned vitrified bricks used as a decorative material in the joints of the bell tower of the Pieve di San Giovanni Maggiore. The parameters derived from the analysis of the masonry construction techniques seem to be only partially helpful for the definition of the roles of the subjects involved in the constructions of this historical period. For example, the analysis of the masonries testifies to the diffused good squaring and roughing of ashlars of medium-large dimensions for the realization of regular facings, probably an indication of the presence of specialized workers of local provenance. The ashlars surface finishing is mainly linked to the characteristics of the raw material (sandstone or limestone) rather than to the masons or clients.

Thirteenth century: the main factor that characterizes this period seems to be the choice of bricks. From a typological point of view, the extremely heterogeneous quality/nature of the brick walls do not allow the proposition of specific categories. However, the use of bricks in historic buildings finds very specific historical and stratigraphic attestations, which lead to the definition of a chronological sequence in its utilisation. In particular, after a first attestation of this material, around the middle of the thirteenth century for the rebuilding of the Convent of San Francesco in Borgo, the bricks were applied immediately after (the second half of the thirteenth century) in the extensive restorations carried out in the two Franciscan convents in Mugello (San Francesco a Borgo and Bosco ai Frati) and in the reconstruction of the bell tower of the Pieve di Borgo San Lorenzo. From that moment, the bricks became part of the building materials of the area, arriving in the second half of the sixteenth century to replace the Pietra Alberese. Therefore, the bricks could have been introduced during the thirteenth century for the convent of San Francesco in Borgo San Lorenzo, built probably with the arrival of Francesco in Mugello (Altieri 2009).

About the workers who produced and realized this brickwork, the morphological similarity between the bell tower of the Pieve di Borgo San Lorenzo (Figs. 1.10 and 1.11), built between 1258 and 1263, and the Lombard-style bell towers present in central and Northern Italy could testify the movement of Lombard workers in Mugello in the thirteenth century. The presence of these workers

Fig. 1.10 The north-east elevation of the bell tower of the Pieve di Borgo San Lorenzo

would have led to the consequent introduction of bricks among the construction materials of the area in the thirteenth century, a phenomenon already widely documented for other sites in Tuscany by numerous published works, which focused on the study of written sources (Salvestrini 2011) and the analysis of material evidence (Bianchi 1996; Frati 2013a, 2013b; Parenti and Quiros Castillo 2000; Pittaluga and Quiros Castillo 1997; Quiros Castillo 1997, 2001).

For this reason, it appears possible that the introduction of this new building material could be related to the arrival of specialized builders from the North and then, given the rapid expansion in its use, become part of the know-how of local workers during the Fourteenth and fifteenth centuries.

As for the stone, it is possible to propose precise considerations from the analysed construction techniques. The example that perhaps best testifies to the use of skilled workers in this period is the Pieve di Sant'Agata del Mugello, where the construction material, namely the Pietra Alberese, is put into place through a very well-executed squaring and surface smoothing of the ashlars; such factors could therefore testify to a remarkable ability in the processing of materials by specialized builders, probably hired by wealthy clients. In other cases, such as in the Pieve di San Gavino Adimari, the restoration operations were carried out using local raw materials, roughly hewn, or split and with a laying tending to regularity. Currently, there are no sources which testify to the use of specialized or local workers and the clients who commissioned these works.

Second half of the sixteenth–early seventeenth century: intense reconstructions took place following the earthquakes of 1542, 1597, and 1611 (Arrighetti and Minutoli 2019). In this

Fig. 1.11 The south elevation of the bell tower of the Pieve di Borgo San Lorenzo

case, the kind of activities provides a clear overview of the workers and clients involved. For most of the studied buildings, the damage mechanisms activated by earthquakes were not followed by constructive moments of particular importance. On the contrary, only reconstructions aimed at solving individual "problems" are attested (for example, fracture reconnections and chain insertions, carried out with recycled materials or mixed raw materials, laid down irregularly). Only in one case the written sources (Baccini 1893; Baldi 1878) testify to the intervention of a specific client, represented by the Medici family, which used local specialized workers for the Pieve di San Gavino Adimari. In this case, the reconstructions carried out on the church and the bell tower highlight a strong desire on the part of the client to give a new splendour to the building under its patronage.

The Pieve di Sant'Agata constitutes a context where intense reconstructions were conducted at the end of the sixteenth century and in the beginning of the seventeenth century. This is testified by written sources (Brunori Cianti 2011) and by the stratigraphic reading (Arrighetti 2015), which point out the presence of two different clients and workers. The first intervention (following the earthquake of 1542) was carried out by the inhabitants of the village of Sant'Agata. They reconnected the cracks and partially rebuilt the church bell tower with recycled materials or mixed raw materials (Fig. 1.12 and see for further details Montaubert).

A second construction phase was commissioned by Pievano Nozzolini, who employed local and specialized workers for the actual rebuilding of part of the church and bell tower, with Pietra Alberese and Pietra Serena (Fig. 1.13 idem for Fig. 1.12). In this case, therefore, operations as a consequence of earthquakes that occurred in 1542 and 1597–1611 differ considerably from each other in terms of the raw materials used and for the processing and finishing of the construction materials. In general, it was the economic resources of the various clients and, consequently, the more or less specialized local workers engaged in the post-earthquake works that influenced the rebuilding work conducted in this chronological period.

Twentieth century: The earthquake of 1919 was the main event that led to intense changes in the structural construction system. Following the event, numerous interventions ranged from the reconnection of the crack patterns to the complete reconstruction of some buildings damaged by the telluric event. Both written sources and archaeological analysis attest to the evident determination of the central authority to restore certain structures to their splendour, carrying out accurate reconstructions through internal and external transformations. This is the case, for example, of the Parish Church of Borgo San Lorenzo, which was almost totally rebuilt, or the Parish churches

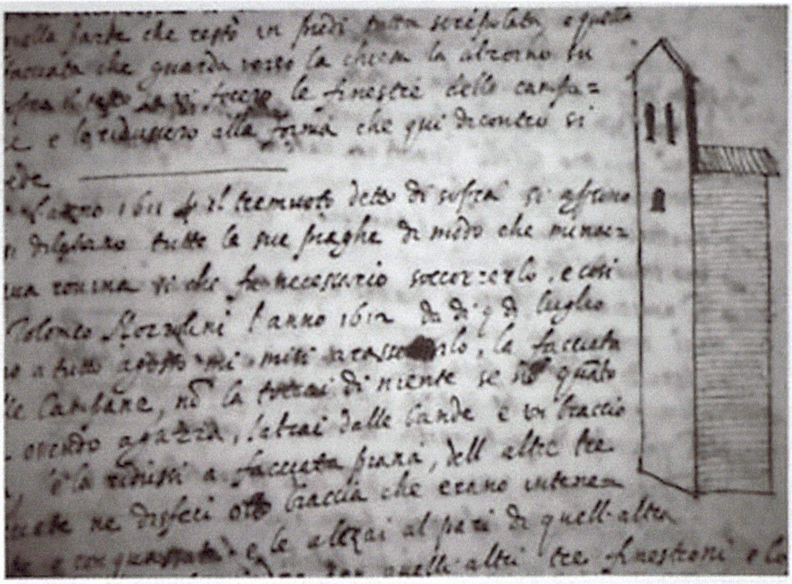

Fig. 1.12 The sketch of the morphology of the bell tower of the Pieve di Sant'Agata del Mugello following the restorations carried out after the earthquake of 1542 by local workers (Brunori Cianti 2011, p. 88)

of San Cassiano in Padule and San Cresci in Valcava, which were demolished to be rebuilt from the foundations (Fig. 1.14).

Some churches, probably because of their isolated position or their decreasing importance in the social dynamics of the period, were only partially restored. Among these are the Pieve di San Gavino Adimari, the Church of Santa Maria a Fabbrica, and the Church of San Bonaventura del Bosco ai Frati.

1.6 The Study of the Bedding Mortars of the Masonries

The characterization of bedding mortars can have considerable potential in archaeoseismology studies. These materials contain much technological information, both in terms of raw materials and the 'recipe' of their realisation. Their study contributes to the characterization of the masonries even when they are realized with the

Fig. 1.13 The Pieve di Sant'Agata in a photo from 2009

Fig. 1.14 The church of San Cresci in Valcava, in a photo of 2009, visibly shows its total rebuilding following the earthquake of 1919. The only portion of the building that survived the earthquake is the bell tower

same stone materials and with similar processing and finishing.

The following information will be given on the nature of mortars and why they can provide relevant information.

Mortars are artificial materials produced by man since ancient times for various functions, including laying stones and bricks for the realization of the masonries, plastering, pasting tiles, or creating decorations (stucco, etc.). There is also an abundance of mortars with special uses (such as filling holes and sealing).

The mortars are obtained by mixing water and aggregates with a binder, capable of giving the mixture cohesion. It is also possible to add organic additives (such as egg yolk, egg white, casein, animal glue, fig milk, nut oil, linseed oil, blood, and resins), natural fibres (for example, straw, animal hair and iron filings synthetic fibres and pigments) to achieve special characteristics.

The aggregate is added to the mixture to reduce shrinkage and increase the final product's strength. Generally, the aggregate consists of sand from river deposits but can also come from beaches and unconsolidated sand outcrops. The mechanical crushing of rocks can also produce an aggregate.

The following categories can be distinguished among the binders: clay binders, gypsum binders, and lime-based binders. In relation to the latter, the most common in the Italian territory are calcic limes obtained by burning limestones, magnesian limes obtained by burning dolostones, hydraulic limes obtained by burning marly limestones, and limes that have become hydraulic by the addition of particular components.

As for the study of these materials, until the early 1980s, the characterization was mainly based on traditional wet chemical analyses (Jedrzejewska 1960; Cliver 1974; Frizot 1977;

Dupas 1981). However, this type of investigation posed problems of interpretation of the analytical data due to the nature of the mortar, consisting of a binder and an aggregate, which are often difficult to separate. Subsequently, it was decided to undertake a mineralogical-petrographic study as if the mortars were a natural compound (rock) (Franchi et al. 1985; Middendorf et al. 2000; Van Balen et al. 2000; Martinet and Quenée 2000; Crisci et al. 2004; Pavia and Caro 2006; Pecchioni et al. 2014).

The petrographic study observes thin sections of the mortar samples under an optical microscope. This allows for the determination of its mineralogical composition, the texture (micritic, microsparitic, sparitic), and structure (homogeneous, lumpy, plague) of the binder and its interaction with the aggregate. It allows the recognition of the presence of materials that give the mixture hydraulicity, such as pozzolana or cocciopesto. These materials, containing Si and Al compounds in amorphous form, react with lime to form hydrated calcium silicate and hydrated calcium and aluminium silicates, determining a setting even in high humidity or submerged environments.

Regarding the aggregate, the observation of thin sections allows for recognition of the mineralogical and petrographic composition of the aggregate (therefore to identify its origin), the granulometry, and the shape (rounding and sphericity) that allow identifying any selection/sieving, natural origin (sand), or artificial (crushed rocks). The aggregate distribution in the paste can indicate the care taken in mixing the paste itself. Based on the quantity of binder with respect to the aggregate, it is possible to distinguish mixtures with a high binder/aggregate ratio (fat mortars) and mixtures with a low binder/aggregate ratio (lean mortars). Observing the thin section also allows the description of the macroporosity, with the appearance of the pores, the distribution and the percentage by volume, and the description of the additions (such as straw), if present. Regarding the raw materials used in the realization of a mortar, this type of study allows information on the supply areas. Different raw materials can be linked to the different construction phases.

In the context of Mugello archaeoseismology, the first application of the petrographic study of the bedding mortars was made on the Parish Church of Sant'Agata as part of an international project entitled *AcRoSs—Archaeoseismology, Restoration, Seismology*. The project involved an interdisciplinary group with researchers from two nations (Italy-France). Between 2018 and 2019, the team developed an innovative research project for the analysis of the building based on the creation of a digital model of the church and the annexed buildings through the integration of products. These products were derived from the application of modern technologies, the design of which integrated historical-archaeological analyses, the three-dimensional survey of the structure with the aid of laser scanners, terrestrial and drone photogrammetry, analyses on construction materials with particular reference to mortars and stone and brick elements, and a series of information obtained through the auscultation of the response building to seismic noise. The results have not yet been published, but a personal communication from the authors confirmed the usefulness of the data provided, especially in differentiating construction phases otherwise not easily recognizable on a macroscopic basis. Other studies of bedding mortars in this area of Tuscany, not connected, however, to studies of archaeoseismology, concern the sixteenth century Medici Fortress of San Piero a Sieve and the tower of the Trebbio Castle (fifteenth–sixteenth century). About the Fortress, the mortars are characterized by very fatty mixtures (binder/aggregate ratio varying from 1/1 to 1/2) and a fine aggregate of unimodal granulometry (Fig. 1.15).

Regarding the binder, the presence of burnt remains of Pietra Alberese indicates this limestone is the raw material for the production of lime (Fig. 1.16) (Fratini et al. 2018). The exceptional mechanical characteristics of these mortars would suggest the possible selection of a variety of Pietra Alberese richer in clay, capable of providing a high hydraulicity (Fratini et al. 1994). The widespread recrystallization

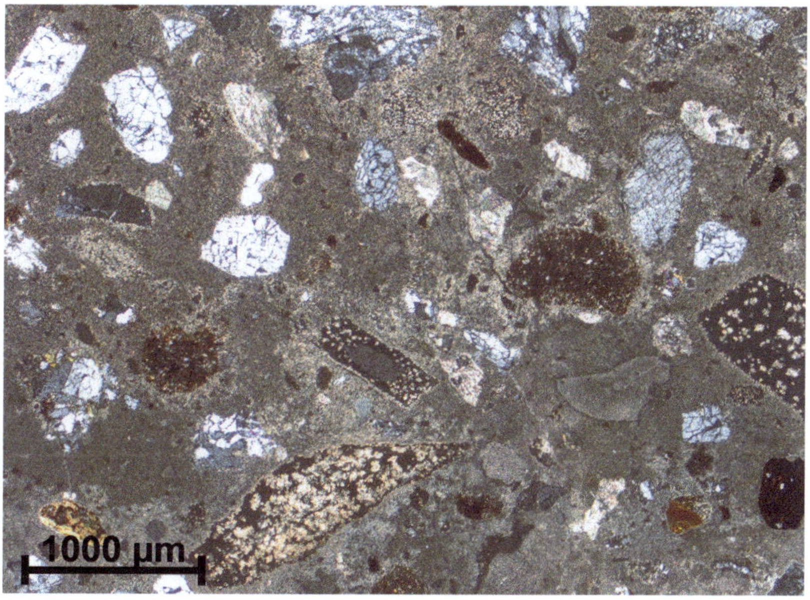

Fig. 1.15 Thin section in polarized transmitted light. It is possible to see the abundant binder and the unimodal grain-size aggregate consisting of single crystals of quartz and feldspar

phenomena suffered by the binder must also be considered to justify the performances of these mortars (Fig. 1.17). Overall, for their petrographic aspect and effective binding properties, these mortars are comparable with those present in Brunelleschi's Dome of the Cathedral of Santa Maria del Fiore in Florence (Barbi et al. 1986; Pecchioni 2012). Different types of mixtures have been identified for the mortars taken from Castello del Trebbio, which are distinguished by the amount of binder (Figs. 1.18 and 1.19) (unpublished data).

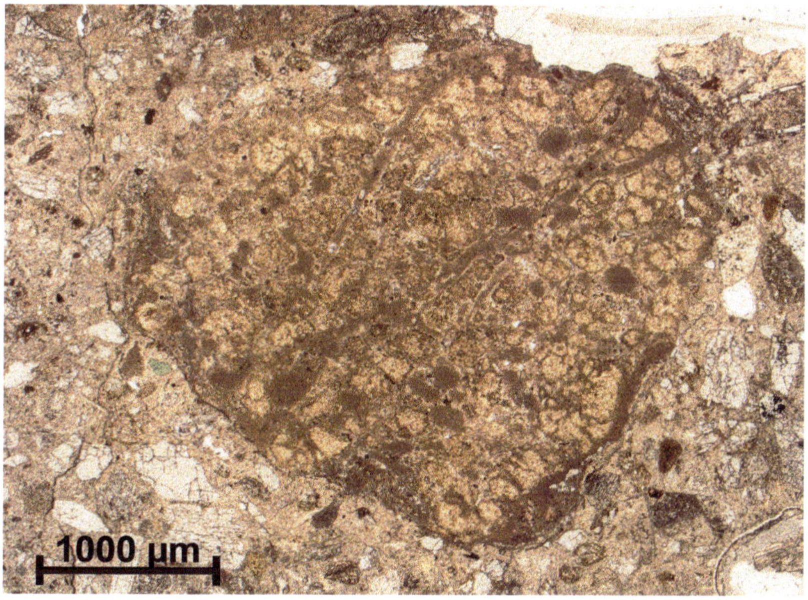

Fig. 1.16 Thin section in polarized transmitted light. Underburnt fragment of Pietra Alberese indicates that this limestone was the raw material for the production of lime

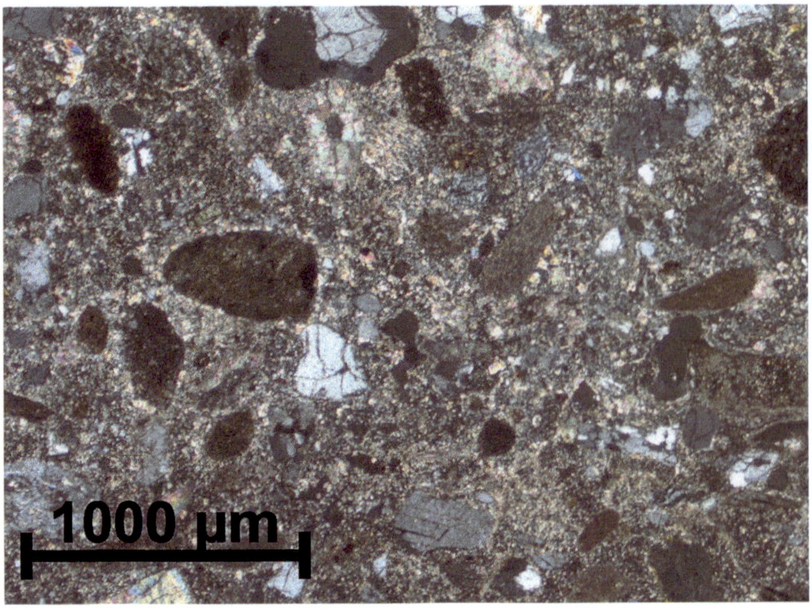

Fig. 1.17 Thin section in polarized transmitted light. The strong recrystallization of the lime binder is evident

Fig. 1.18 Thin section in polarized transmitted light. It is possible to see the aggregate of bimodal grain size consisting of arenaceous and limestone fragments, single crystals of quartz, feldspar, and sparitic calcite (Castello del Trebbio)

Fig. 1.19 Thin section in polarized transmitted light. It is possible to see the abundant binder and the unimodal grain-size aggregate consisting of single crystals of quartz, feldspar, and sparitic calcite, arenaceous fragments, and rare cocciopesto (Castello del Trebbio)

1.7 Conclusions

Clients, workers and raw materials are the three key elements behind the main choices made in Mugello in the various historical periods for buildings and activities from restoration to reconstruction. The results that emerged from the research indicate that these factors influenced the construction of buildings in a certain place and historical period. Mostly of religious origin, the clients have often resorted to specialized workers, probably already present and operating in the area during the above-mentioned times. Instead, the use of non-native workers seems to be documented around the middle of the thirteenth century for the sole construction of the Pieve di Borgo San Lorenzo bell tower, where the building model shows architectural characteristics related to the Lombard style. The adoption of raw materials appears to be linked to purely technical and practical factors. The availability and exploitation of outcrops of specific construction materials and the introduction of more promising building technology (the introduction of brick which along the three centuries replaces the Pietra Alberese) guided the choices made by clients and workers.

Post-seismic operations deserve a separate discussion, where a sort of differentiation seems to emerge. In the sixteenth century, the buildings testify to a double reality: the hasty restorations carried out a few days after a given seismic event, made by the inhabitants themselves to solve the most evident damage caused by the earthquake (such as cracks reconnections and partial reconstruction of roofs) with rather approximate construction techniques, and more specialized choices undertaken until a century after the events and linked to the requests of specific clients (such as the Medici family and some parish priests). In this second case, the accuracy in the construction materials' selection, processing, and equipment denotes the presence of specialized workers who were probably local residents. In the twentieth century, the situation changed again: following the earthquake of 1919, the most prestigious buildings in city centres underwent major restorations while the isolated architectural complexes remained abandoned or made safe through temporary measures which became definitive, making them vulnerable to structural hazards.

References

Altieri A (2009) Notizie del Borgo San Lorenzo in Mugello e suo territorio raccolte da Valentino Felice Mannucci nel 1742–1743. Protagon Editori Toscani. Colle Val d'Elsa

Arrighetti A (2015) L'archeosismologia in architettura: Per un manuale. Firenze University Press, Firenze

Arrighetti A (2022) Analisi e documentazione archeologica dell'edilizia storica in aree a rischio sismico: dall'atlante dei tipi costruttivi murari all'impiego delle moderne tecnologie di rilievo. Archeologia e Calcolatori 33(1):201–218

Arrighetti A, Minutoli G (2019) A multidisciplinary approach to document and analyze seismic protection techniques in Mugello from the middle ages to early modern time. Ann Geophys 62(3):SE335. https://doi.org/10.4401/ag-7991

Baccini G (1893) La pieve di San Gavino Adimari presso Barberino di Mugello: Memoria del P. Pompeo Franchi. In: Giotto. Bollettino Storico Artistico Letterario del Mugello, Firenze

Baldi F (1878) La pieve di San Gavino Adimari e Benedetto Varchi: ricordi storici, Firenze

Barbi L, Leggeri B, Vasarri V, Franchi R, Fratini F and Manganelli Del Fà C (1986) Indagine sperimentale sui materiali costituenti la Cupola di Santa Maria del Fiore. Firenze. Università di Firenze-Facoltà di Architettura Dipartimento di Costruzioni, 1: 5–7

Bessac JC (1986) L'outillage traditionnel du tailleur de pierre de l'antiquité à nos jours. CNRS, Parigi

Bianchi G (1996) Magistri d'Europa: eventi, relazioni, strutture della migrazione di artisti e costruttori dai laghi lombardi, Maestri costruttori lombardi nei cantieri della Toscana centro-meridionale (secoli XII-XV). In: Della Torre S, Mannoni T, Pracchi V (eds) Indizi documentari ed evidenze materiali. Proceedings of the Congress, Como 23–26 ottobre 1996, pp 155–166

Brogiolo GP, Cagnana A (2012) Archeologia dell'architettura: Metodi e interpretazioni. All'Insegna del Giglio, Firenze

Brunori Cianti L (2011) Sant'Agata: una santa, una pieve, una comunità del Mugello, Firenze

Carmignani L, Lazzarotto L (2004) Carta Geologica della Toscana 1:250.000. Direzione delle politiche territoriali ed ambientali–Servizio Geologico d'Italia: Regione Toscana, Firenze, Italy

Cipriani N, Fratini F, Nebbiai M, Sartori R (2005) L'Arenaria di Monte Senario: caratteristiche composizionali, tecniche e confronto con la Pietra Serena. Arkos-Scienza e Restauro dell'Architettura 6(9):37–44

Cliver EB (1974) Tests for analysis of mortar samples. In: Bulletin of Association for Preservation Technology (APT), 6th edn. Association for Preservation Technology (APT), New York, pp 68–73

Crisci GM, Franzini M, Lezzerini M, Mannoni T, Riccardi MP (2004) Ancient mortars and their binder. Periodico di Mineralogia, Special issue 3A, showcase of the Italian research in applied petrology. Ed. Bardi Roma 73:259–268

Dupas M (1981) L'analyse des mortiers et enduits des peintures murales et des bâtiments anciens. In: Proceedings of the ICCROM symposium mortars, cements and grouts used in the conservation of historic building. ICCROM, Rome, pp 281–295

Ferrari G, Molin D (1986) Campo macrosismico del terremoto del 13 giugno 1542. Atti del IV Convegno. Roma 29–31 ottobre 1985. CNR-GNGTS1, pp 373–386

Fiorani D (1996a) Tecniche costruttive murarie medievali: il Lazio meridionale, Roma

Fiorani D (1996b) Le tecniche costruttive murarie medievali del Basso Lazio. Metodo e percorsi di una ricerca. In: Della Torre S (ed) Storia delle tecniche murarie e tutela del costruito. Esperienze e questioni di metodo, Milano, pp 97–112

Franchi R, Fratini F, Manganelli Del Fà C (1985) Caratterizzazione degli Intonaci mediante l'impiego di tecniche Mineralogico-Petrografiche. In: Proceedings of the Congress Scienza e Beni Culturali: L'Intonaco: Storia, Cultura e Tecnologia. Bressanone, Ed Libreria Progetto Padova, pp 167–183

Frati M (2013a) Verso un atlante delle murature ad Empoli: la mensiocronologia del laterizio. Milliarium 10:126–135

Frati M (2013b) Tracce lombarde nella Toscana protoromanica. In: Proceedings of the International Congress, Architettura dell'XI secolo nell'Italia del Nord. Storiografia e nuove ricerche. Pavia, 8–10 aprile 2010, pp 253–270

Fratini F, Giovannini P, Manganelli Del Fà C (1994) La Pietra da calce a Firenze: ricerca e caratterizzazione dei materiali per la produzione di 'calcina forte' e 'calcina dolce'. In: Proceedings of the Congress Scienza e Beni Culturali 10: Bilanci e Prospettive Bressanone. 5–8 Luglio 1994, Padova. Libreria Progetto Editore, pp 189–199

Fratini F, Pecchioni E, Cantisani E, Rescic S, Vettori S (2014) Pietra Serena: the stone of the renaissance. In: Pereira D, Marker BR, Kramar S, Cooper BJ, Schouenborg BE (eds) Global heritage stone: towards international recognition of building and ornamental stones, geological society, vol 407. Special Publications, London, pp 173–186

Fratini F, Arrighetti A, Cantisani E, Pecchioni E (2018) I materiali da costruzione della Fortezza di San Martino a San Piero a Sieve (Toscana, Italia). In: Marotta A, Spallone R (eds) Defensive architecture of the Mediterranean, vol VII. Politecnico di Torino, pp 381–388

Fratini F, Cantisani E, Pecchioni E, Pandeli E, Vettori S (2020) Pietra Alberese: building material and stone for lime in the Florentine territory (Tuscany, Italy). Heritage 3:1520–1538

Fratini F, Rescic S, Pittaluga D (2022) Serpentinite and ophicalcite in the architecture of eastern Liguria and as decoration of Tuscan religious buildings. Resour

Policy 75:102505. https://doi.org/10.1016/j.resourpol.2021.102505
Frizot M (1977) Stucs de Gaule et des provinces romaines: motives et techniques. Université de Dijon, Centre de recherches sur les techniques gréco-romaines, Dijon
Jedrzejewska H (1960) Old mortars in Poland: a new method of investigation. Stud Conserv 5:132–138
Mannoni T (2005) Archeologia della produzione architettonica. Le tecniche costruttive. Arqueología de la Arquitectura 4:11–19
Martinet G, Quenée B (2000) Proposal for a valuable methodology for studying ancient mortars. In: Paisley (ed) Proceedings of the international RILEM workshop historic mortars: characteristics and tests. Chapman & Hall, London, pp 81–91
Middendorf B, Baronio G, Callebaut K, Hughes JJ (2000) Chemical mineralogical and physical-mechanical investigations of old mortars. In: Proceedings of the international RILEM-workshop. Historic mortars: characteristics and tests. Chapman & Hall, London, pp 53–61
Moretti I, Stopani R (1974) Architettura romanica religiosa nel contado fiorentino, Firenze
Parenti R (1987) Una proposta di classificazione tipologica delle murature postclassiche. In: Proceedings of the Congress Scienza e Beni Culturali: Conoscenze e sviluppi teorici per la conservazione di sistemi tradizionali in muratura. Bressanone 23–26 giugno 1987, pp 49–61
Parenti R, Quiros Castillo JA (2000) La produzione dei mattoni della Toscana medievale (XII-XVI secolo): un tentativo di sintesi. In: Boucheron P, Thebert Y, Broise H (eds) Proceedings of the International Congress La brique antique et médiévale: production et commercialisation d'un materiau. Paris, 16–18 Nov 1995, pp 1–17
Pavia S, Caro S (2006) Lime mortars for masonry repairs: analytical science and laboratory testing versus practical experience. International seminar theory and practice in conservation, Lisbon:493–500
Pecchioni E (2012) Analisi mineralogico-petrografiche dei campioni del sondaggio S2 della Cupola del Brunelleschi. Il monitoraggio delle grandi fabbriche storiche- 60 anni di monitoraggio della cupola del Brunelleschi. Firenze, Mandragora, pp 142–145
Pecchioni E, Fratini F, Cantisani E (2014) Atlas of the ancient mortars in thin section under optical microscope. Nardini Ed, Firenze
Pierotti P (2003) Manuale di Sismografia Storica. Lunigiana e Garfagnana, Pisa
Pierotti P, Ulivieri D (2001) Culture sismiche locali, Pisa
Pittaluga D, Quiros Castillo JA (1997) Mensiocronologie dei laterizi della Liguria e della Toscana: due esperienze a confronto. In: Proceedings of the Ist National Congress Archeologia Medievale. Pisa, 29–31 maggio 1997, pp 460–463
Quiros Castillo JA (1997) La mensiocronologia dei laterizi della Toscana: problematiche e prospettive di ricerca. Archeologia dell'Architettura 2:159–166
Quiros Castillo JA (2001) Mensiocronologia e produzione di laterizi in Toscana in età medievale. In: Proceedings of the Congress I laterizi in età medievale. Dalla produzione a cantiere. Roma, 4–5 giugno 1998, pp 31–44
Salvestrini F (2011) Eremitismo–Cenobitismo. La realtà di Santa Maria di Vallombrosa in età medievale. In: Bertocci S, Parrinello S (eds) Proceedings of the Congress Architettura eremitica. Sistemi progettuali e paesaggi culturali. Vallombrosa, 24–25 settembre 2011, pp 33–39
Van Balen K, Toumbakari EE, Blanco Varela MT, Aguilera J, Puertas F, Sabbioni C, Zappia G, Riontino C, Gobbi G (2000) Procedure for a mortar type identification: a proposal. In: Proceedings of the international RILEM-workshop historic mortars: characteristics and tests, Paisley. Chapman & Hall, London, pp 61–79

Reading the Earthquake of 1542 in the Walls of the Church of Sant'Agata del Mugello: The Need for a Multidisciplinary Approach

2

Arnaud Montabert

Abstract

Historical buildings are a unique opportunity to record and study the effects of past earthquakes at a given place. Seismic traces are recorded as damage or construction repairs. These direct and indirect traces' diversity, rarity, and complexity require a multidisciplinary approach.

An innovative methodology is defined to improve knowledge of local historical seismicity. Such a methodology is based on a multidisciplinary approach combining: analysis of historical sources, stratigraphic analysis, and structural analysis of a historical building. The church of Sant'Agata del Mugello (Tuscany, Italy) is considered a case study. The stratigraphic analysis is performed by identifying the repairs using the RECAP methodology. Eighty repair units using 13 building techniques are identified in the church. The identified repairs are associated with unknown events, earthquakes, or routine reconstructions. When post-earthquake reconstructions are found, damage mechanisms are associated with them. Thirteen constructive phases of the church have been traced by combining stratigraphic analysis and historical sources. This work is focused on the 1542 seismic event (epicentral macroseismic intensity 9, deduced magnitude 6.0), which resulted in the collapse of the upper part of the bell tower and the two lateral chapels as well as the overturning of the front wall and the two lateral walls of the nave. The contributions and limits of each domain are discussed.

Keywords

Historical earthquakes · Methodology · Seismic traces · Stratigraphic analysis · Multidisciplinary approach · Post-earthquake reconstructions

2.1 Introduction

Because of their suddenness and unpredictability, earthquakes remain among the most complex and dangerous natural phenomena. To protect the population, their prediction remains one of the great challenges of the seismological community. We distinguish between short-term forecasting and long-term forecasting.

For short-term forecasting, unfortunately, there is currently no way of predicting precisely where, when, and how strong an earthquake will occur. Indeed, the precursory signs of an earthquake are not always identifiable and interpretable. Long-term forecasting consists of evaluating a region's seismic hazard, i.e. the probability of

A. Montabert (✉)
Laboratoire de Géologie de l'Ecole normale supérieure, PSL University, CNRS, U.M.R.8538, Paris, France

CNRS, LMPS–Laboratoire de Mécanique Paris-Saclay, Université Paris-Saclay, ENS Paris-Saclay, CentraleSupélec, Gif-sur-Yvette, France

L. Pecchioli (ed.), *Archaeoseismology*, Natural Science in Archaeology,
https://doi.org/10.1007/978-3-031-28303-1_2

an earthquake occurring in a given area over a given period (50 years, 500 years, etc.). Understandably, quantifying historical earthquakes becomes critical in evaluating the seismic hazard on such a time scale. During the last decades, palaeoseismology, archaeoseismology, and historical seismicity allowed for identifying alternative sources of information.

These sources can be very diverse: historical texts, destruction layers, geological markers, and evidence of damage in archaeological remains. Among them, historical buildings witnessed ancient earthquakes as "stone seismometers" recorded on their walls as structural disorders and repairs. The diversity, rarity, and complexity of the traces left by ancient earthquakes require a multidisciplinary approach.

To illustrate this need, it is proposed to study the earthquake of 1542 in the basin of Mugello by studying the building history of the mediaeval church of Sant'Agata del Mugello through the eye of four disciplines: historical seismology, history, building archaeology, and earthquake engineering.

The reader can find an exhaustive analysis of the building evolution of the church's construction in previous studies (Arrighetti 2015; Montabert et al. 2020). However, the purpose of this article is:

- To propose a methodology to highlight the effects of historical earthquakes on a historical building
- To demonstrate the need for a multidisciplinary approach
- To discuss the strengths, limitations, and complementarity of each field

2.2 Description of the Case Study: Sant'Agata del Mugello

Sant'Agata del Mugello is a Romanesque church (Fig. 2.1 and see for further details Arrighetti et al.) located in the Mugello basin 30 km North of the city of Florence (Tuscany, Italy).

The rectangular nave (15.7 m × 29.3 m, Fig. 2.1) draws its singularity by its unique wooden frame (Fig. 2.2); its walls are about 0.80 m thick. The central apse (15.7 m × 29.3 m, Fig. 2.3) and the two squared lateral chapels (4.4 m × 4.4 m and 4.5 m × 4.5 m for the North and South chapel, respectively) are overhung by a vault system; their wall thicknesses range from 0.40 m to 0.70 m. The bell tower (Figs. 2.1 and 2.3) is 21 m high. Its plan is almost square (6.3 m × 6.0 m), and the masonry walls are 1.10 m thick. Its construction has been established at least since the tenth century. The church is, therefore, a witness to the many

Fig. 2.1 External view of Sant'Agata del Mugello (CC By SA)

Fig. 2.2 Internal view of Sant'Agata del Mugello (photo credits: Maria Lancieri)

earthquakes that have taken place in the Mugello basin.

2.3 Seismotectonic Context and Historical Seismicity

The Mugello area is one of the intramontane basin along the NE-dipping Etrurian Fault System. The basin is 25 km long. The sedimentary composition of the basin is complex due to two consecutive phases of infilling. The first fluvial-lacustrine phase occurred during the late Pliocene—early Pleistocene period. An alluvial phase occurred during the Pleistocene (Benvenuti 2003). A compressive regime in Late Pliocene-Early Pleistocene related to the contemporary Northern Apennine active thrust (50 km northeast from the basin) developed the basin. The basin was then affected by NE-SW extension and normal faulting when the compressive regime ceased around the Early-Middle Pleistocene period (Sani et al. 2009) concerning the opening of the northern Tyrrhenian Sea.

Two large antithetic normal fault systems border the Mugello basin: the Ronta Fault System (RFS) and the Sieve Fault System (SFS). Two transverse normal fault systems flank the western and eastern parts: the Scarperia Fault System and the Vicchio Fault System (Fig. 2.4). The seismogenic potential of these fault systems is a matter of debate. Some elements are presented to demonstrate the complexity of the RFS and the SFS.

2.3.1 The Sieve Fault System

The Sieve Fault System is a NE-dipping antithetic normal fault bounding the southwestern margin of the basin (Fig. 2.4) with a slip rate of about 0.7 mm/year (Delle Donne 2005). Martini and Sagri (1993) and Benvenuti (2003) interpret the fault system as the master fault of the basin for two main reasons: the Sieve Fault system seems to be in the continuity of the Etrurian fault; the authors argue that the control of the basin highlighted by the asymmetry of the basin characterised by a deeper bedrock in the southwestern part of the basin (sediment thickness about 600 m). Based on this hypothesis, the Ronta Fault System (Fig. 2.4), located in the North, is interpreted as a subordinate fault system. Based on these arguments, the DISS group (2018) (DISS 2018) subdivided the southwestern basin margin into a north-western (DISS ID: ITIS087) and a south-eastern (DISS-ID: ITCS037) fault segment (two yellow boxes in Fig. 2.4).

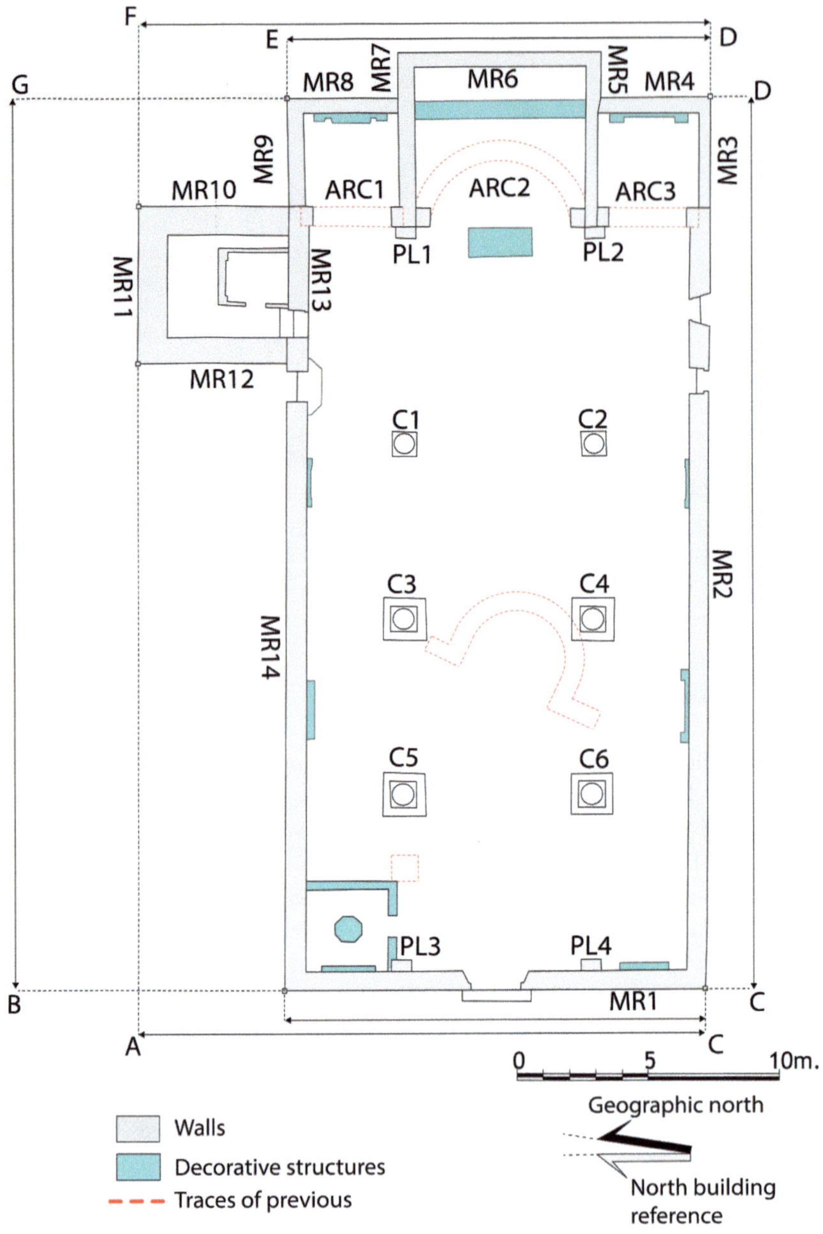

Fig. 2.3 Plan of the church with the location of technical groups: MR for the wall, PL for the pilaster, C for the column, and ARC for the arch. Labelled lines (AC, CD, . . .) are related to interpreted orthophotos presented after that (Montabert et al. 2020)

2.3.2 The Ronta Fault System

The RFS is formed of SW-dipping antithetic normal faults bounding the northeastern margin of the basin (Fig. 2.4) with a slip rate of about 0.5 mm/year (Delle Donne 2005). This fault system has attracted attention in recent years. It is regarded by Bonini et al. (2016) as the most likely source of the most significant earthquakes that struck the Mugello area based on two arguments. First, the RFS depicts a lateral continuity (Fig. 2.4) contrary to the SFS, the Vicchio Fault System, and the Scarperia Fault System. Second, the area exhibits clear morphological evidence. The digital elevation is depicted. Moreover, the morphological features of a section of the RFS are steeper than sections of the other fault system.

Instrumental seismicity has recently allowed the recognition of another normal fault system in the north-western Mugello basin margin activated

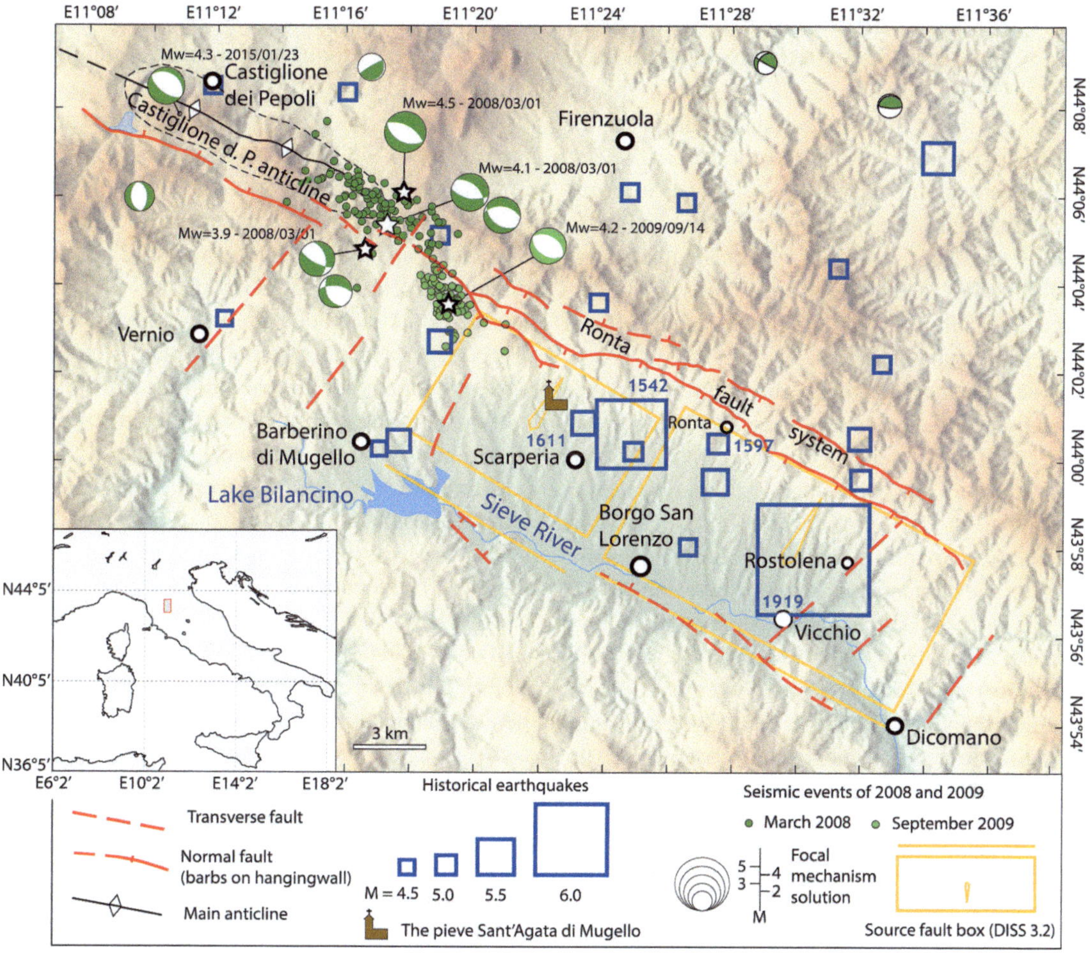

Fig. 2.4 Seismotectonic context of the Mugello area (adapted from Bonini et al. 2016 and location of the church of Sant'Agata del Mugello). The active faults are compiled from various sources: Delle Donne (2005), Sani et al. (2009). Historical earthquakes are from the CPTI15 catalogue (https://doi.org/10.6092/ingv.it-cpti15). The main shocks of 2008 and 2009 are from Amato et al. (2008)

in 2008 and interpreted as a steep NNE-dipping seismogenic normal fault (Amato et al. 2008; Ripepe et al. 2008) and the 2009 seismic events interpreted as a steep SSW-dipping seismogenic normal fault as a northwards extension of the Ronta Fault System. The Scarperia Fault System separates this fault system from the Ronta Fault System. The activity of the north-western part of the basin has been recently confirmed by the occurrence of an Mw = 4.6 earthquake that struck the area between Barberino di Mugello and Scarperia on 9 December 2019, 100 years after the strongest event in 1919.

2.4 Historical Seismicity

A "rather high seismicity characterised the basin". The strongest known historical earthquakes occurred in 1542 (Mmw = 6.0) and 1919 (Mmw = 6.4). The two earthquakes induced extensive damage over the entire province; based on damage intensity, 1542 is located in the north-west part of the basin, while 1919 is in the south-east. More minor earthquakes occurred but were not reported in the entire basin (Rovida 2016). This is the case for 1597 (Mmw = 5.3) and 1611 (Mmw = 5.1) earthquakes.

2.4.1 Contributions and Limitations of the Seismotectonic Analysis

Although none of the candidate faults can be clearly associated with each historical earthquake, the seismotectonic context gives crucial information. First, the Mugello basin's seismic activity is mainly characterised by the Ronta Fault System and the Sieve Fault System. The role of the Vicchio Fault System and the Scarperia Fault System remains unclear.

The distance between the church and the potential candidate fault varies according to the expected fault system. In the case of the SFS, the distance may vary between 0 (the hypocentre belongs to the NE-dipping fault) and 17 km (the southeastern extremity of the SFS). For the same reason, the distance may vary between 0 and 18 km for the RFS. Distances between the church and the VFS may be between 13 km (south extremity) and 15 km (south extremity 0 is not permitted since the Vicchio Fault System is supposed to be a sub-vertical fault system). For the same reason, we expect distances between 9 km and 10 km in the case of the Vicchio Fault System. If the northern extension of the RFS, characterised by the recent instrumental seismicity of 2008–2009, is involved, distances between 5 km and 15 km are expected.

The recent seismic crisis of 2008 and 2009 may produce at least moderate seismicity of Mw at about 4. It is not excluded that this zone could generate stronger earthquakes related to the historical seismicity, which highlights stronger magnitude (the Mmw = 6.01542 and the Mmw = 6.41919 earthquake).

2.4.2 Using Historical Information from Archives

Historical records, where available, provide localised information. However, studying historical seismicity requires an exhaustive search of historical records. These new data allow an absolute dating of the identified action (damage, repairs, etc.). However, it is necessary to carefully distinguish primary and secondary sources since they play an essential role in interpreting the impact of the earthquake and the resulting uncertainty (Albini et al. 2017).

One of the first largest earthquakes occurred on 6 June 1542, with an estimated epicentral macroseismic intensity of 9 and an estimated magnitude of 6.0. The earthquake is reported in 81 historical records, among which 27 are from direct sources (CFTI5—Guidoboni et al. 2019). Scarperia is located in the epicentral area (Fig. 2.4). In Sant'Agata del Mugello's particular case, the damage is reported in four main historical writings. Descriptions are different from one historical source to another. A letter (primary source) dated 23 June 1542, kept in the Biblioteca Nazionale Marciana di Venezia reported damage, such as collapsed buildings, injuries, and fires in the area. The locality of Sant'Agata, composed of around 100 houses, is described as collapsed with 11 fatalities. The church is described as destroyed except for a part of the bell tower described with a bell still attached at the top of the tower:

> Santa Agata quale era honorevel vila de più di cento fuochi e li erano molte bellissime case sta peggio del altre, tutta si trova rovinata e così un bel tempio che li era e il campanile tutto sfassciato, sopra del quale li è rimasto la campana quale sta in bilico ne si vede come né chi la sustenti talché pare cosa stupenda e meravigliosa e lì sono morti XI.

A second primary document originates from the Archivio di Stato di Firenze provides detailed information about 1742 collapsed buildings and 113 people. The particular case of Sant'Agata is described as a collapsed church with six injuries: "S[an]ta Agata Villa rovinata la chiesa, e mortovi 6 persone".

A third primary document from the Archivio di Stato di Pistoia gives some quantitative information about 1500 collapsed buildings. The church is again described as collapsed: "Sancta Agata tutta la chiesa e la casa rovinata; morti 6".

Finally, a primary document from the Archivio Parrocchiale dei Santi Jacopo e Antonio di Fivizzano described the church as being destroyed with six fatalities:

> Castelgrosso e Santagheta ruvinata la chiesa e morto persone 6.

The "Libro Campione V" is a detailed secondary source of information for the 1542 earthquake has been used. It is a unique document describing with a high degree of accuracy and maintenance works done in the church from 1596 until 1630 (Cianti 2011). The logbook was initiated in 1596 by the priest Raffalello Sbaccheri (1595–1605) (Cianti 2011) and continued from 1608 to 1630 by the priest Tolomeo Nozzolini (1569–1643). However, this source is not referenced in the Italian databases (CPTI, CFTI). Indeed, the church is a typology of structure too specific to be used in the macroseismic studies (MSK or EMS-98 scale).

In 1608, Tolomeo Nozzolini found the church Sant'Agata highly damaged. He interpreted the damage as the consequence of the 1542 earthquake, and, although he did not witness the earthquake, collected first-hand information about its effects. Nozzolini described the tilting of the facade and learned from a villager that the bell tower was higher before the 1542 earthquake. He noticed that the villagers rebuilt only the southern wall. Two windows were opened, and the roof was covered. A sketch was drawn by Nozzolini when he found the bell tower in 1608 (Fig. 2.5 and see for further details Arrighetti et al.).

Archives provide crucial but contradictory information on the effects of the 1542 event on the church of Sant'Agata del Mugello. The texts describe two different levels of damage. On the one hand, the primary sources describe an almost collapsed church. On the other hand, Nozzolini's logbook describes damage located only in the bell tower and the church's main façade.

2.5 Stratigraphic Analysis Based on the RECAP Method

The stratigraphic analysis is based on the RECAP method "REconstruire Après un tremblement de terre: Expériences antiques et innovations à Pompéi"; Rebuilding after an earthquake. Ancient experiments and innovations in Pompeii, ANR-14-CE31-0005, 2015–2019 (http://recap.huma-num.fr), developed as part of the study of post-earthquake reconstruction in Pompeii. The RECAP method focuses on the identification of architectural repairs and comprises three distinctive components:

- Identifying building techniques and unit repairs
- Postulating the damage
- Proposing relative chronological relations between repairs.

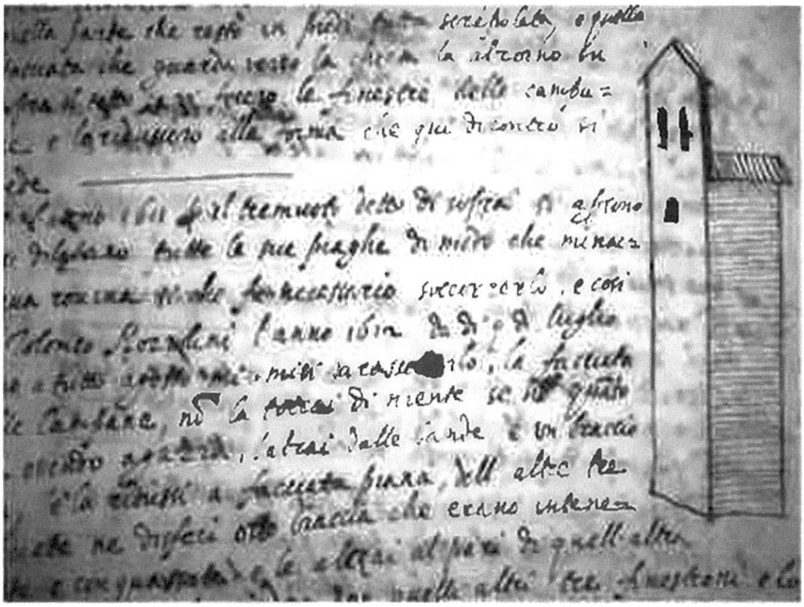

Fig. 2.5 Sketch of the bell tower proposed by Tolomeo Nozzolini in 1608 (Cianti 2011)

The methodology has been applied to the church's internal and external walls (Montabert et al. 2020). Since we focus on the 1542 earthquake, we only show results for sections AC and DF (Fig. 2.3). Each building technique is described in terms of morphology, materials, stone dimensions, and layout of the materials. As a second step, the nature of the damage is determined according to the following disorders: inclinations, deformations; fractures, cracks; supposed collapses. In the case of post-earthquake reconstruction, a failure mechanism can thus be identified depending on the nature of the damage. Because repairs are not necessarily associated with natural disasters but it could be explained by building expansion/reconfiguration, the origin of the damage is discussed, and then the probability (very low, low, strong, extreme). The third step is the description of the nature of the repairs. The report can be done in written form or by filling in a certain number of criteria (such as crack filling, reconstruction, and buttress construction), the dimension of the repair (length, width, height, and thickness), the type of technique used (such as masonry and metallic ties), and the full description of the material used. The fourth and final step is the identification of chronological relations, which are twofold: the chronology of the construction technique applied in the repair is defined concerning the other construction techniques present in the building inventory in the ACoR database (or another reference, indicating, using a code, the types of construction techniques defined); a chronology between the identified repairs into the OPUR database is also done. Thus, a repair can be posterior, anterior, or contemporary to one or more types of construction techniques. By establishing these relationships, it is possible to develop a statistical approach to identify the most damaged kinds of techniques and those most commonly chosen for repairs. In the case of the Sant'Agata del Mugello church study, the data recorded in the OPUR database is available at the following link (https://github.com/MArnaud/OPUR_SantAgata/).

2.6 Identifying Building Techniques

Sixteen building techniques are placed in the church based on the description of the materials' morphology, type of materials, stone dimensions, and layout. The materials used are mainly limestone (Pietra Albarese and Formazione di Silano), sandstone (Pietra Serena), serpentinite (Marmo di Monte Verdi), and bricks. An example of such a description is illustrated in Fig. 2.6. The second step is to distinguish the spatial distribution of these techniques; as shown in Fig. 2.7, 80 construction units are then identified.

Once the different construction units have been distinctly identified, a relative chronology is established between these techniques, as proposed in Fig. 2.8. The same work is presented in Fig. 2.9.

The stratigraphic analysis reveals that the corners of the facade have been restored (repair units 67, 68, 70, and 71 in green in Fig. 2.8). This repair reinforces the link between the facade and the nave. In the case of the nave, repair unit 46 confirms the reconstruction of the southern wall, as reported by Nozzolini. In the same way, the infilling of the cracks (repair units 29 and 30) is still visible on the west wall (MR 12 in Fig. 2.8). The two lateral chapels are also an interesting case. Although no documentation directly described potential damage in this part of the church, a discordance is yet observed between technique TCM10 (in green in Fig. 2.9) and TCM16 (in brown in Fig. 2.9). This is interpreted as a repair (repair units 14 and 21 in Fig. 2.9).

In the case of Sant'Agata del Mugello, the stratigraphic analysis allowed:

- A chronological relationship between the different repair units to be established
- The confirmation of damage or repair operations described in texts (destruction of the upper part of the bell tower, tilting of the upper parts of the nave)
- Clarification of the descriptions of the historical texts. The church was not destroyed by the earthquake of 1542; it was primarily preserved

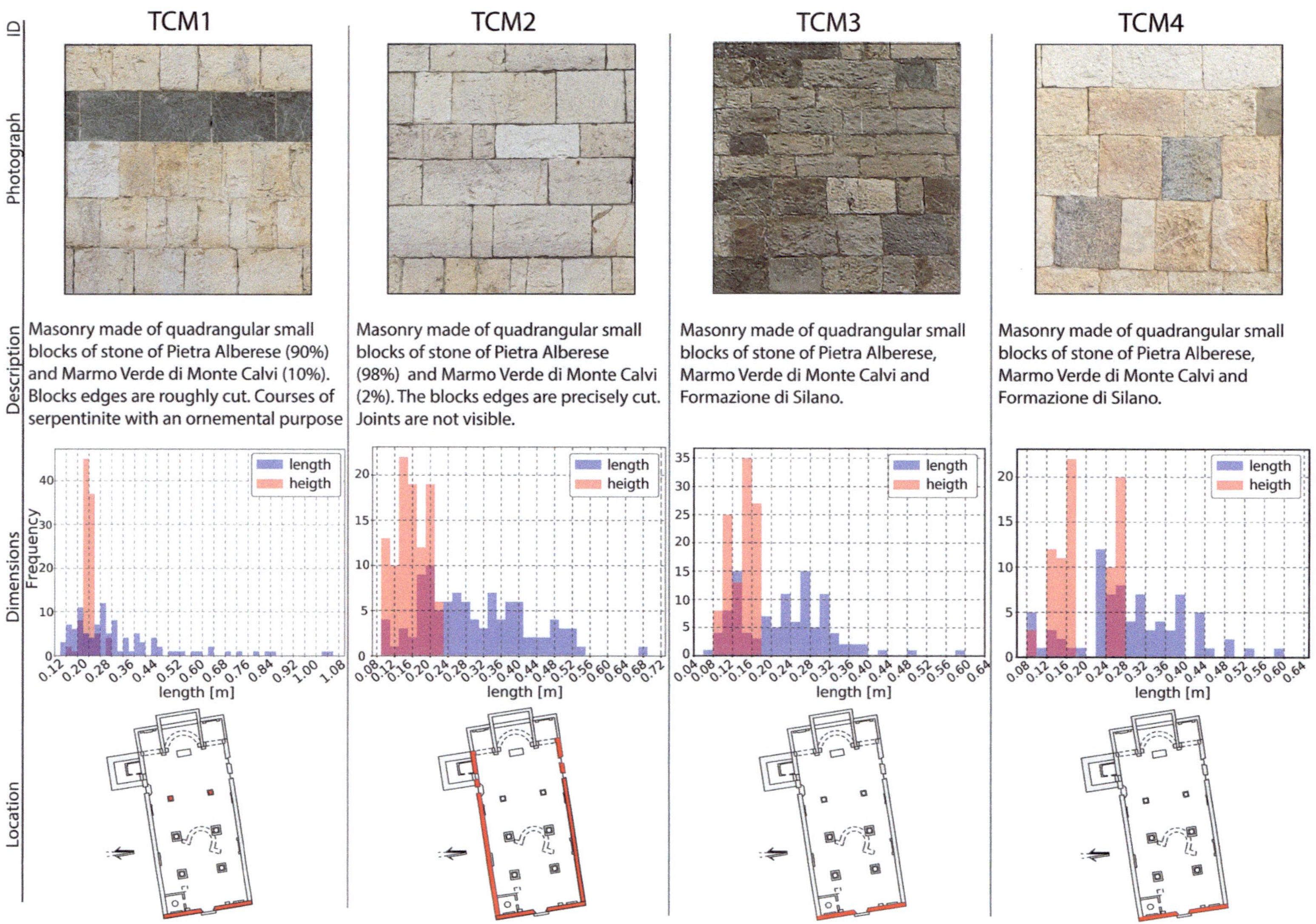

Fig. 2.6 Presentation of the technique of construction TCM1, TCM2, TCM3, and TCM4 found in the church (Montabert et al. 2020)

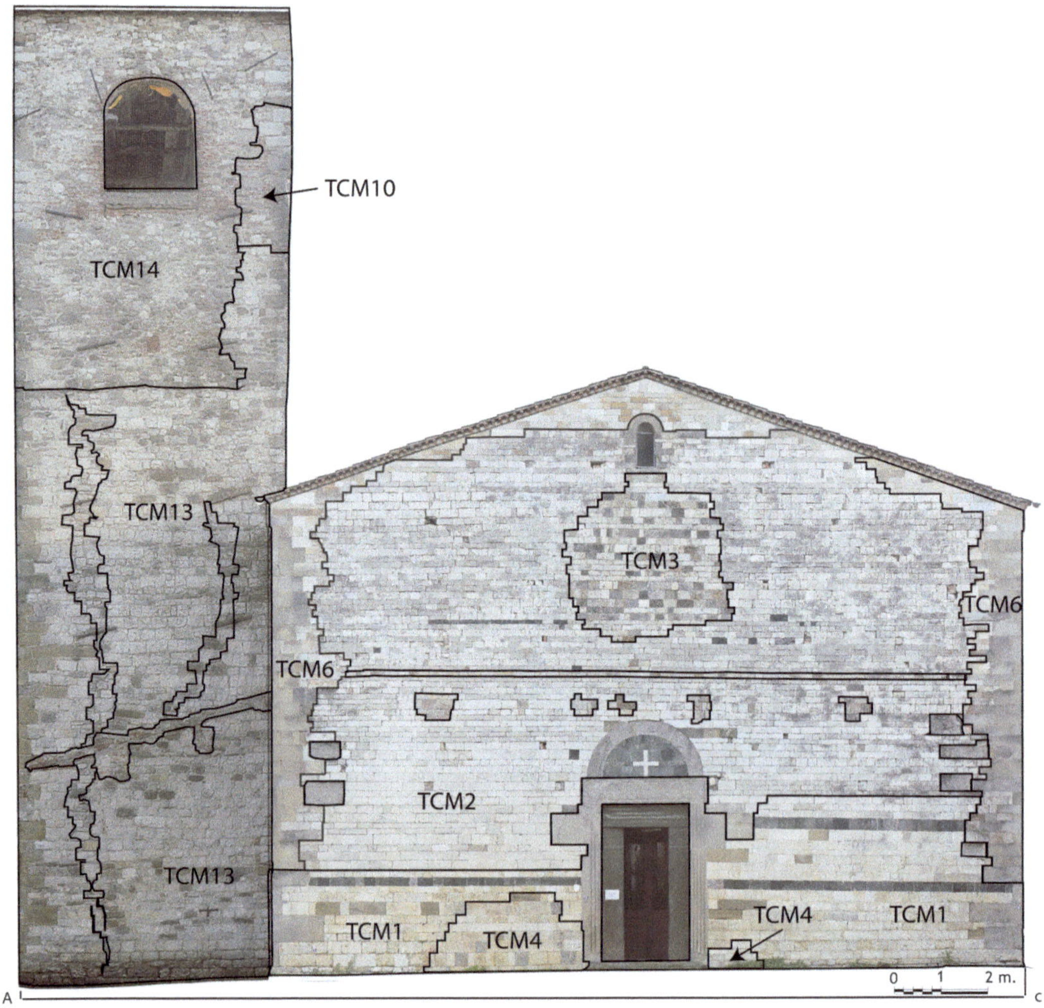

Fig. 2.7 Boundaries of building techniques identified for the church facade

- The missing information from the texts, such as the collapse of the two lateral chapels, to be resolved.

2.7 Identifying Failure Mechanisms from Indirect Traces

The final step is identifying the mechanisms of building failure induced by the 1542 earthquake. The reconstruction of the building over the centuries does not make direct observation of damage possible. It is, therefore, necessary to interpret the identified repairs.

The identification of failure mechanisms is directly inspired by the damage observed in the field during a postseismic survey. First, these mechanisms can be observed in terms of three scales: at the scale of the brick, the masonry panel, or the whole building.

At the brick level, either the repair is non-existent and can therefore be observed, or it consists of using mortar to repair the joint or consolidate the brick. The masonry panel uses mortar to redo the joints and fill the cracks. Some elements of the wall are replaced (such as bricks). At the building scale, failure mechanisms can be grouped into three families, as illustrated in Fig. 2.10.

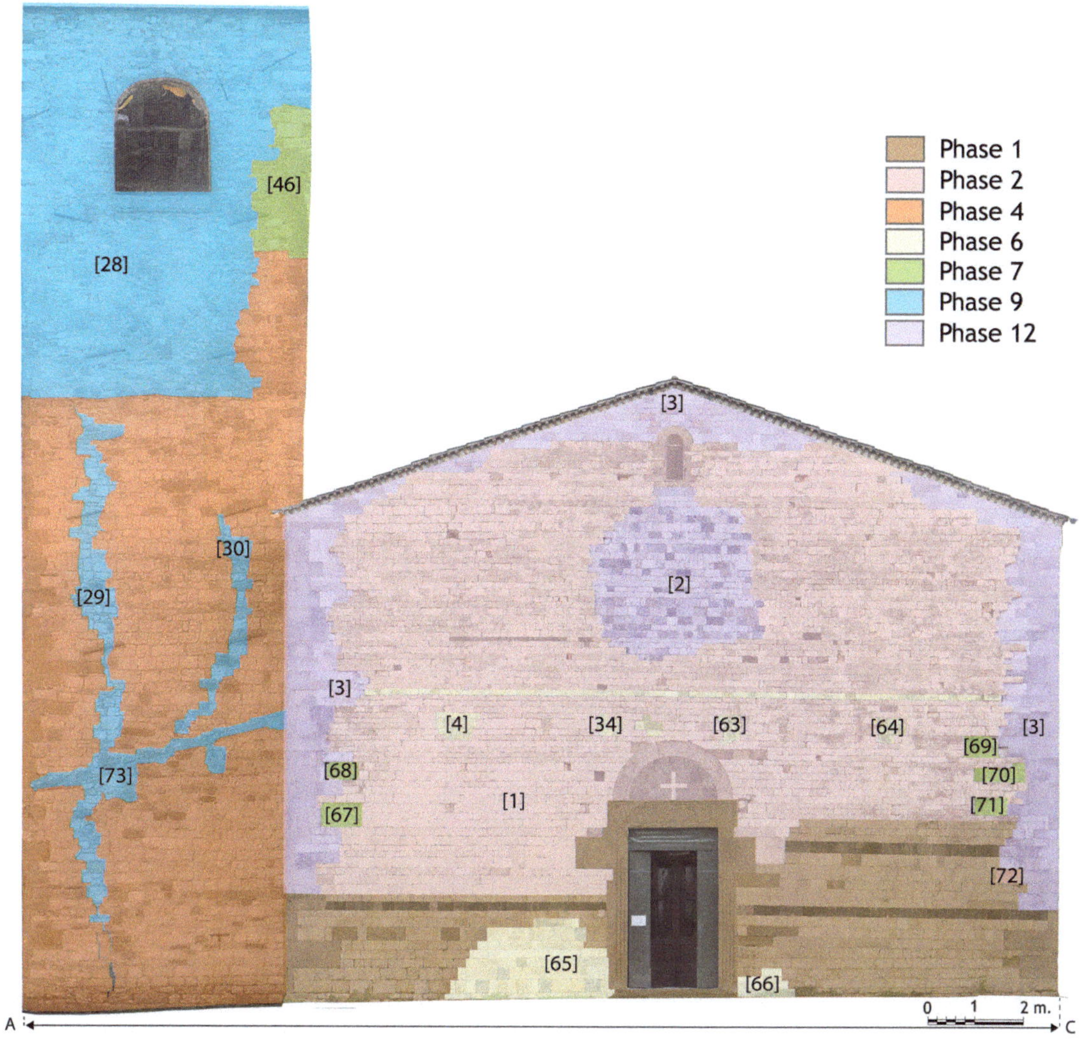

Fig. 2.8 Stratigraphy of the front wall (MR1, external west-facing of the church and MR12, external west-facing of the bell tower) showing phase 1, phase 2, phase 7, phase 8, and phase 13. Phase 1 corresponds to the original construction of the current Romanesque church. Phase 2 is identified as a repair of phase 1 (Montabert et al. 2020)

There are many alternatives for repairing the same type of damage. Let us consider an overturning with two side wings (Fig. 2.11i). The first reconstruction strategy is to build up the collapsed wall (Fig. 2.12b). A second strategy is to add a corner chainage to the wall reconstruction to link the facade to the side walls and avoid a facade collapse (Fig. 2.12c). A brick chainage can also connect the building walls and reinforce the connection between the walls (Fig. 2.12d). Buttresses are sometimes added to prevent further tilting (Fig. 2.12e). Finally, all these seismic techniques can be combined (Fig. 2.12f).

In the case of Sant'Agata del Mugello, repair units [26] and [28] (Figs. 2.8 and 2.9) show a collapse of the upper part of the bell tower and vertical cracking (repairs [29] and [30]). The inclination that could be observed in the facade and the walls of the nave points to a vertical tilting of these walls (Fig. 2.11). The repair of the two

Fig. 2.9 Stratigraphy of the external east-facing of the church (MR4, MR6, MR8), including the bell tower (on the right), the central apse, and both lateral chapels. Phase 3, phase 4, phase 5, phase 7, phase 8, and phase 10 are identified (Montabert et al. 2020)

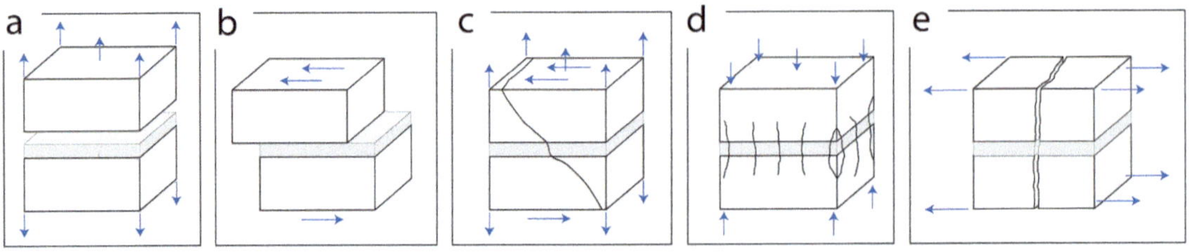

Fig. 2.10 Failure modes at the two-block assemblage scale. (**a**) Block-mortar bond tensile failure, (**b**) Block-mortar bond shear sliding, (**c**) Diagonal masonry cracking, (**d**) Masonry crushing, (**e**) Block and mortar tensile cracking (Montabert 2021)

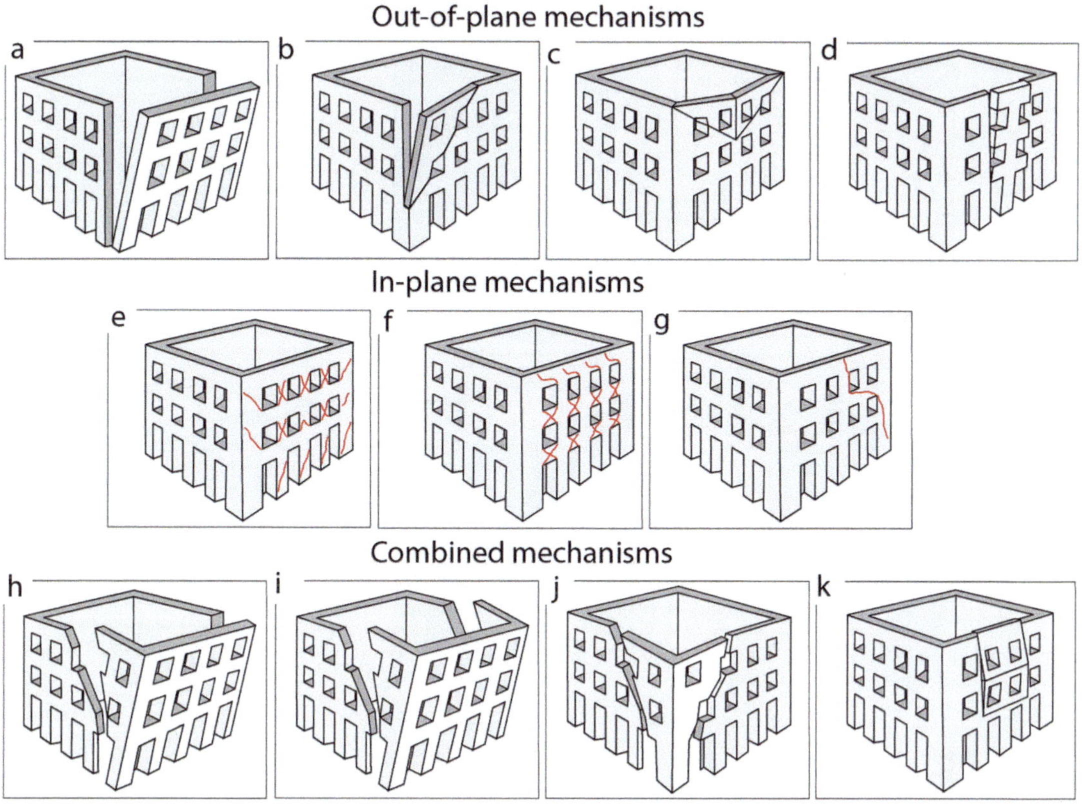

Fig. 2.11 List of failure modes. (**a**) Vertical overturning, (**b**) Partial overturning, (**c**) Horizontal arch, (**d**) Vertical strip overturning, (**e**) Cracks in piers, (**f**) Cracks in spandrels, (**g**) Transverse cracks, (**h**) Overturning with one side wing, (**i**) Overturning with two side wings, (**j**) Corner failure, (**k**) Vertical arch (Montabert 2021)

lateral chapels (repairs [14] and [21]) shown in Fig. 2.9 highlights a collapse of the corner of the two chapels.

2.8 Synthesis of a Multidisciplinary Approach

The earthquake of 1542 is studied through the church of Sant'Agata del Mugello via four interdisciplinary approaches: the analysis of the seismotectonic context, the analysis of archives, the stratigraphic analysis of walls, and structural analysis.

Each approach has provided additional information for a better understanding of earthquake effects. Two faults are identified on the historical seismicity side: the Ronta Fault System and the Sieve Fault System. The historical texts describe the almost total destruction of the building and its repairs. This state of destruction is clarified by the stratigraphic analysis, which shows damage limited to the bell tower, the side chapels, and the inclination of the nave walls. The historical texts nevertheless allow us to accurately date the repair units identified by the stratigraphic analysis.

Furthermore, the stratigraphic analysis provides additional information, such as the damage to the side chapels. Finally, the structural analysis of the repair units allows us to conclude the seismic response of the building during the 1542 earthquake based on postseismic observation of modern earthquakes. Out-of-plane and in-plane mechanisms were identified: vertical cracking of the bell tower, the collapse of the

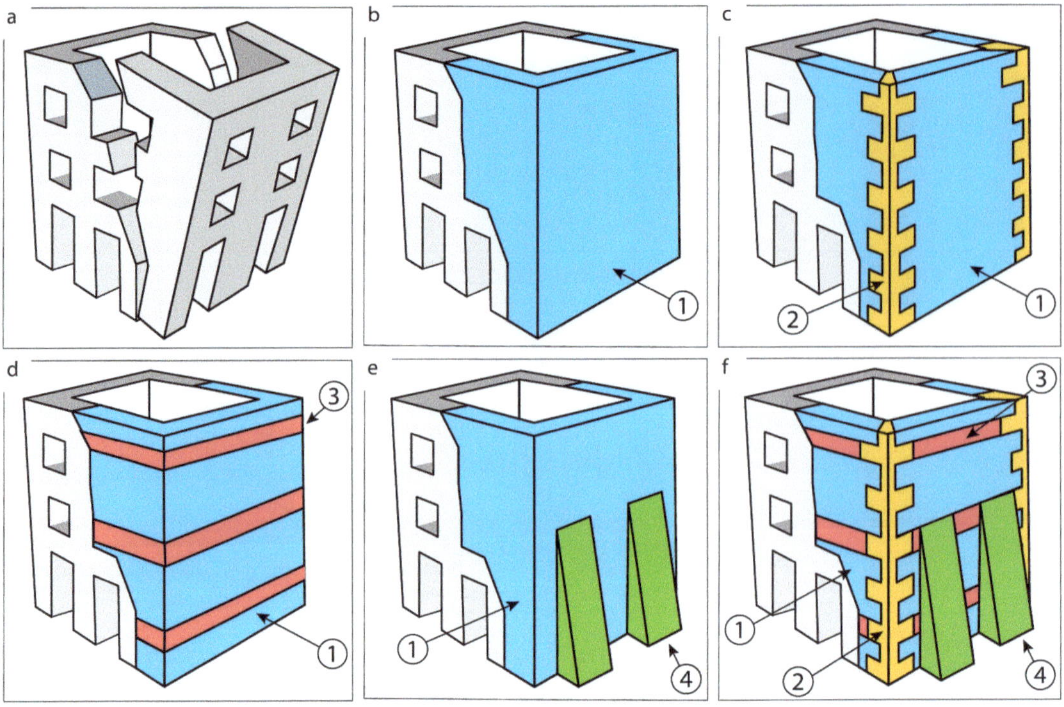

Fig. 2.12 Repair strategies associated with overturning with two side wings (**a**). (**b**) Reconstruction strategy (element 1 in blue). (**c**) Corner chainage technique (element 2 in orange). (**d**) Brick (element 3 in red). (**e**) Buttresses (element 4 in green). (**f**) Combined repair and reinforcement techniques

upper part, tilting of the nave walls, and the corners of the side chapels.

This intense collaboration between the four disciplines is the basis for analysing this historical earthquake. The results of each discipline complement, converge, or sometimes diverge. Above all, they demonstrate the importance of joint analyses between the disciplines.

Acknowledgments I thank my colleagues who actively participated in this work: Andrea Arrighetti, Julien Clément, Hélène Dessales, Maria Lancieri, and Hélène Lyon-Caen. The work greatly benefited from various sources of support: Cecilia Ciuccarelli, who helped in finding the historical sources; Emma Cantisani and Fabio Fratini, who shared results of their ongoing research on mortar analysis; Giovanni Minutoli, Tommaso Francucci, Marco Repole, and Pierpaolo Lagani and the students of the restoration course (2018–2019) from Università degli Studi di Firenze who participated in the photogrammetry acquisition and the field data acquisitions; Raffaella Leporini who provided fruitful comments on archaeological analysis; the students of the Archaeology of Buildings' course at the University of Siena (2018–19), coordinated by Andrea Arrighetti and Giovanna Bianchi, who provided support in archaeological analysis. We also thank Agnes Tricoche (LabEx TransferS) for her collaboration on the OPUR database. The authors warmly thank the community of Sant'Agata del Mugello for their moral and logistic support, particularly the mayor Federico Ignesti, Don Gianluca Mozzi, Marco Casati, Filippo Bellandi, Paolo Gucci, and Anna Bambi. This work was funded by IRSN, ENS, and the programme TelluS-ALEAS of INSU-CNRS.

References

Albini P, Rovida A, Scotti O, Lyon-Caen H (2017) Large eighteenth–nineteenth century earthquakes in western gulf of Corinth with reappraised size and location. Bull Seismol Soc Am 107:1663–1687

Amato A, Antonioli A, Chiaraluce L, De Gori P, Di Stefano R, Doumaz F et al (2008) Rapporto tecnico-scientifico sulla sequenza sismica del mugello del 1 marzo 2008. Instituto Nazionale di Geofisica e Vulcanologia 30

Arrighetti A (2015) Constructive and seismic history of the Sant'Agata del Mugello Parish church. (reuso, Éd.), 64–70

Benvenuti M (2003) Facies analysis and tectonic significance of lacustrine fan-deltaic successions in the pliocene–pleistocene mugello basin, Central Italy. Sediment Geol 157(3–4):197–234

Bonini M, Corti G, Delle Donne D, Sani F, Piccardi L, Vannucci G, Genco R, Martelli L, Ripepe M (2016) Seismic sources and stress transfer interaction among axial normal faults and external thrust fronts in the northern apennines (italy): a working hypothesis based on the 1916–1920 time–space cluster of earthquakes. Tectonophysics 680:67–89

Cianti LB (2011) Sant'agata: una santa, una pieve, una comunit'a del mugello

Delle Donne D (2005) Tettonica attiva dell'appennino settentrionale nel settore compreso tra l'appennino pistoiese e l'alta val tiberina. unpublished phd thesis, 144p

Diss WG (2018) Version 3.2.1: A compilation of potential sources for earthquakes larger than m 5.5 in italy and surrounding areas. diss

Guidoboni E, Ferrari G, Tarabusi G, Sgattoni G, Comastri A, Mariotti D et al (2019) CFTI5med, the new release of the catalogue of strong earthquakes in Italy and in the Mediterranean area. Sci Data 6(1):80

Martini I, Sagri M (1993) Tectono-sedimentary characteristics of late miocene-quaternary extensional basins of the northern Apennines, Italy. Earth Sci Rev 34(3):197–233

Montabert A (2021) Characterizing ground motion of historical earthquakes: study of Sant'Agata del Mugello combining building archaeology, earthquake engineering and seismology. Université Paris Sciences et Lettres, p 40

Montabert A, Dessales H, Arrighetti A, Clément J, Lancieri M, Lyon-Caen H (2020) Tracing the seismic history of Sant'Agata del Mugello (Italy, Tuscany) through a cross-disciplinary approach. J Archaeol Sci Rep 33:102440

Ripepe M, Marchetti E, Delle Donne D, Coli M, Genco R, Ulivieri G, Colò, L, Lacanna G (2008) Dipartimento di scienze della terra di firenze, centro di competenza del dipartimento della protezione civile della presidenza del consiglio dei ministri. 48

Rovida A (2016) Catalogo parametrico dei terremoti italiani (cpti15).

Sani F, Bonini M, Piccardi L, Vannucci G, Delle Donne D, Benvenuti M, Moratti G, Corti G, Montanari D, Sedda L, Tanini C (2009) Late Pliocene–quaternary evolution of outermost hinterland basins of the northern Apennines (Italy), and their relevance to active tectonics. Tectonophysics 476:336–356

The Partial Collapse of the Basilica of Maxentius in Late Antiquity: An Evaluation of Possible Destruction Causes

3

Luise Albrecht and Marina Döring-Williams

Abstract

It is still commonly considered that the main collapse of the Basilica of Maxentius in Rome occurred in the Middle Ages, caused by one of the destructive earthquakes of those centuries. In this paper, the possible causes of destruction—for which, however, evidence can already be found in Late Antiquity—are systematically evaluated also in terms of co-seismically induced damages. Thus, it could be demonstrated that the building's structure exhibited considerable deficiencies already during the construction phase. At the western wall of the central nave in particular, the lateral forces from vault thrust and earth pressure were apparently miscalculated in an almost reckless manner. Moreover, there is indication that static settlement effects have been seriously underestimated, especially in relation to the demolished and backfilled predecessor structures, the *horrea piperataria*, in the subsoil of the construction site of the basilica. These structural weaknesses were so severe that even small tremors or static shifts could have triggered the collapse, even more a major seismic event. And indeed, some horizontal fracture edges can be observed, particularly in the vaults, indicating a sudden lateral load impact as in an earthquake rather than a slow, steady settlement. Thus, the identified construction and settlement related deficiencies should have led to the basilica's (partial) collapse at the first relevant change of specified loads. The earthquake of 443 AD is proposed and discussed here as the closest 'suspicious' event.

Keywords

Rome · Late antiquity · Earthquake · Seismic event · Building archaeology · Structural miscalculation · Construction processes

L. Albrecht (✉)
Fakultät VI Planen Bauen Umwelt, Institut für Architektur, Technische Universität Berlin, Berlin, Germany

Institut für Kunstgeschichte, Bauforschung und Denkmalpflege,Technische Universität Wien, Vienna, Austria
e-mail: luise.albrecht@tuwien.ac.at

M. Döring-Williams
Institut für Kunstgeschichte, Bauforschung und Denkmalpflege,Technische Universität Wien, Vienna, Austria
e-mail: doering-williams@tuwien.ac.at

3.1 The Basilica of Maxentius and 'Its' Earthquakes

The late antique Basilica of Maxentius is located in the heart of Rome, near the Forum Romanum and Colosseum. With its dimensions of almost 100 m in the longitudinal direction and almost 60 m in the transverse direction this imperial audience hall[1] is widely thought to be among

L. Pecchioli (ed.), *Archaeoseismology*, Natural Science in Archaeology,
https://doi.org/10.1007/978-3-031-28303-1_3

Fig. 3.1 Ruin of the Basilica of Maxentius, illustration by Étienne Du Pérac, first published in 1575: debris field after the collapse of the central nave and southern aisle with a view towards the barrel vaults of the preserved northern aisle and the only partially destroyed northern apse with a probably dramatised inscenation on the collapsed vault fragments in the foreground. Illustration from: Du Pérac (1621)

the largest vaulted spaces in antiquity. Its erection not only entailed demolition of a large number of the existing structures and massive urban development changes, but also required radical adjustments to the topography in the area of the building site and its surroundings.

Today, only about a third of the original construction still exists: mainly the northern side aisle with its three characteristic huge barrel vaults, parts of the northern nave wall and the adjacent northern part of the narthex, and furthermore, truncated walls and pillars of southern side aisle, central nave, and western apse (Fig. 3.1).

The restoration of the Basilica of Maxentius (1998–2003) on the occasion of the *Grande Giubileo del 2000*, headed by the Soprintendenza alle Antichità di Roma, provided an opportunity to launch a comprehensive collaborative project.[2] The authors were involved in the joint project on conducting building-archaeology investigations, which was organised by the Technische Universität Berlin and funded by the Fritz Thyssen Foundation, Cologne. The project headed by J. Cramer (TU Berlin) and M. Döring-Williams (TU Berlin/TU Wien) included the first detailed geodetic survey of the construction, together with manual documentation and complex analysis.

With the present work, the authors aim to contribute to the earthquake discussion from the perspective of building archaeology, which primarily takes into account in situ observations of the historical architecture in question and aspects of building materials and construction techniques, as well as analyses of the processes on the ancient building site. Finally, they hope that their conclusions will further advance the discourse on the cause(s) of the basilica's collapse through expert assessments of the geology, dynamics, and structural engineering.

In most of the literature concerned, it is assumed on the basis of historical considerations that the main collapse of the Basilica of

[1] Ziemssen (2011: 227)

[2] Giavarini (2005: 9–10)

Maxentius was caused by an earthquake in the Middle Ages, focusing on the horrendous disasters of 847 and 1349.[3] However, to date no written source could definitely be linked to one of these events, and neither do the archaeological or the structural finds prove a clear connection with any of them.

In recent years, initial structural analyses were undertaken by Alejandra Albuerne.[4] She was concerned, among other things, with the central nave cross vaults and with the situation after the collapse in the *horrea piperataria* in the subsoil of the basilica, where the cracked vaults that penetrated the floor of the basilica ended up. These *horrea* are the predecessor structures, an imperial-era market complex, which were partially filled in and built over when the basilica was constructed. Based on evidence of lateral load damage to the *horrea* structure, Albuerne confirms an earthquake impact and dates it vaguely to after the construction of the Basilica of Maxentius, but without considering other impairment factors in this area.[5]

3.2 Collapse of Parts of the Basilica as Early as the Fifth Century?

So there is a long list of authors who see a seismic event as the trigger for the collapse of the Basilica of Maxentius.[6] Consistently, the majority of researchers base their suggestions for the year of the crash on the dates of the known and particularly devastating earthquake catastrophes of the Middle Ages in Rome. However, in 2010, evidence was found that an important part of the building may have collapsed much earlier than previously thought.

The reason for this assumption is the variously observed reuse of very characteristic building material and even entire construction components (wall and vault fragments as coming from the debris of the Maxentian basilica itself) within portions of secondary securing, repair, and conversion works in the complex, which are archaeologically and structurally dated to Late Antiquity.[7]

As with the entire complex of the Basilica of Maxentius, only the northern area of the narthex is preserved. According to archaeological investigations, constructions on top of the narthex roof were carried out in two phases '*nell'arco del V secolo*'—in the course of the fifth century AD.[8] A subsequent development with various rooms on the roof of the narthex of a prestigious public building is somewhat surprising. All the more so when these rooms were interconnected with walls and pillars in the northern surrounding street (Figs. 3.2 and 3.3), which was therefore no longer passable.[9] In addition, it is noticeable that already these late antique structures were apparently intentionally limited on the part of the narthex roof still preserved today. There is no evidence for a former continuation of the masonry beyond the broken edge to the south; on the contrary, some kind of safe distance seems to have been maintained to the fracture zone (see top view narthex in Fig. 3.2). This interpretation is further strengthened by the observation of a new roof

[3] Albrecht and Döring-Williams (2010: 197) with a summary of previous research positions: Note 9, p. 215

[4] For example, Albuerne (2010, 2016); Albuerne and Williams (2017)

[5] The *horrea piperataria* have not yet been studied in detail. There is a lack of comprehensive investigations and reliable reconstructions of the former market area and its buildings as well as a lack of knowledge about the techniques, procedures, and static consequences of their (partial) demolition and backfilling during the Maxentian construction site.

[6] see Note 2

[7] Albrecht and Döring-Williams (2010: 205–212); Fabiani and Coccia (2003: 38–39)

[8] 'I livelli di frequentazione individuati all'interno degli ambienti erano costituiti da semplici battuti di terra: soltanto in alcuni tratti si conservavano i resti di un massetto di malta. Questi strati hanno restituito consistenti contesti ceramici databili nell'arco del V secolo'. Fabiani and Coccia (2003: 38–39)

[9] See the floor plan drawings of the *domus* by Guglielmo Gatti in: Buzzetti and Virgili (1985: 314–320); the plan of the *intercapedine* in Barosso (1940: pl. XVa); photos of the narthex roof during excavation in 1999 in Fabiani and Coccia (2003: 49–51); selected photos during demolition in 1932 in Amici (2005: 61–62); section through *domus*, *intercapedine*, and narthex in Döring-Williams and Albrecht (2020: 374)

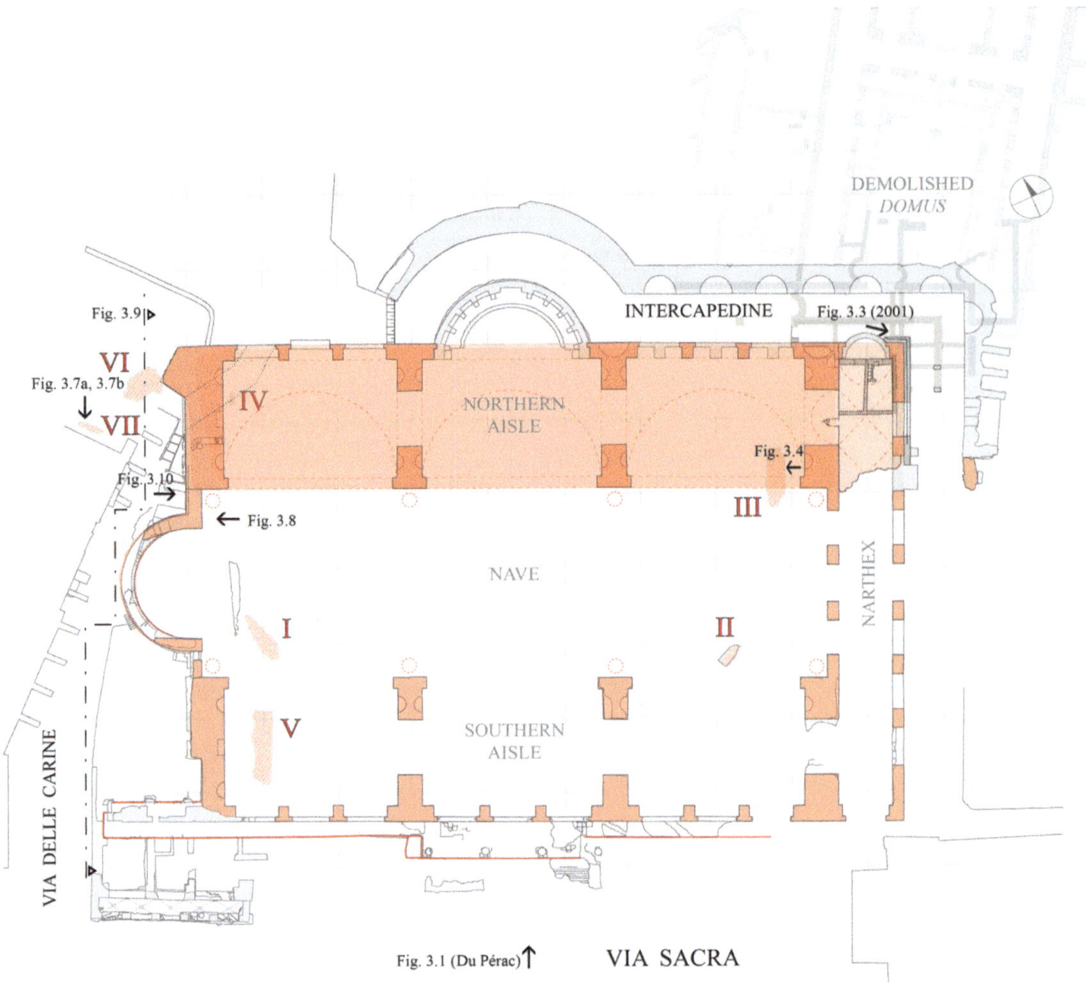

Fig. 3.2 Basilica of Maxentius, Rome. Plan showing present state. Maxentian walls, red; preserved Maxentian vaults, light-red; Post-Maxentian, grey; denominated (vault) fragments (I–VII) and references to figures. Grid lengths: 10 m × 10 m. Plan: Authors, including details of (excavation) plans by Barosso, Fabiani, and Coccia, Gatti (see Note 9)

drainage of the same construction period (Fig. 3.3). Remarkably, it also starts at this edge, thus apparently taking only the mentioned area of the roof into account. The rest of the roof seems to have simply no longer existed or at least to have been unfit for further development.[10] In summary, from a building archaeological perspective it seems that these fifth century building activities had to respond to already destroyed or lost building sections and to react to subsequently abandoned areas.

[10] Albrecht and Döring-Williams (2010: 208–212)

If the authors were to stick to the tradition of specifically looking for the causes of the observed destruction of the Basilica of Maxentius in the earthquakes, the favourite for 'nell'arco del V secolo' (see Note 7) would distinctly be the seismic event of the year 443.[11] It must have been a terrible natural disaster, so terrible '*ut primae eius aedes aedificiaque corruerint*'—that its [Rome's] main temples and buildings collapsed.[12]

[11] Albrecht and Döring-Williams (2010: 213)

[12] Paulus Diaconus, Historia Romana XIII, 16 (eighth century) cited after Galli and Molin (2014: 1281)

Fig. 3.3 Secondary drainage system of the northeastern narthex roof built in the course of the fifth century AD. Photo: A. Pauleck, G. Styhler-Aydin

According to Galli and Molin, this earthquake may have damaged the Colosseum, San Pietro in Vincoli, and San Clemente's buildings, among others.[13] The still not conclusively identified '*portica nova*' are also described as affected.[14] Another candidate in the period in question might be the earthquake of 508, which apparently caused considerable damage again in the immediate vicinity of the Basilica of Maxentius—as, for example, certainly to the Colosseum and possibly to the Templum Pacis.[15] However, at the current state of the archaeological investigations, the suggestion about the basilica's collapse in 443 must be considered more plausible (see Note 7).

[13] Galli and Molin (2014, Table 1, p. 1294)

[14] Fasti Vindobonenses Posteriores (fourth–seventh century) cited after Galli and Molin (2014: 1280–1281)

[15] Galli and Molin (2014, Table 1, p. 1294)

Fig. 3.4 Vault fragment from the nave, 'negative' of the roof cladding, Maxentian construction phase: imprints and remains of the antique roof tiles (*tegulae*) laid in mortar above a stratum of *cocciopesto*. Photo: L. Albrecht

3.3 Evaluation of Possible Destruction Causes

As building archaeologists, it is also imperative for the authors to rule out possible triggers other than earthquakes for the Basilica of Maxentius, not least because very precise dating approaches are in question here.

Mainly two studies are relevant for the compilation of possible alternative causes of collapse, namely Galadini et al. (2006), where a feasibility matrix with seven categories was developed, and Galadini et al. (2018), with the adaptation of causes. Now we have applied some of these categories in a modified and focused way again to the Basilica of Maxentius. This allowed us to look more closely and more systematically at and discuss the feasibility of the respective causes of the collapse.

3.3.1 Unfinished Structure/Roofing

Massive damage to buildings can occur if they remain unfinished for a certain period of time, i.e., if the components that are supposed to protect them from the weather, such as the roof, but also roof overhangs, eaves, cornices, gutters or façade and roof coverings, are missing. Recently, evidence could be presented that the structural work and most probably also the roof covering (Fig. 3.4) were completed already at the Maxentian construction site.[16] Some peripheral areas of the complex as well as major parts of the interior decoration and exterior wall plaster, apparently, were not.[17] Recent archaeological and building archaeological findings suggest that, contrary to the previous opinion, the northern apse was added only between the second half of the fourth century and the beginning of the fifth century. In this second phase, work continued in the exterior areas and on the (interior) decoration, but at least in the *intercapedine* in the north, the façade plastering and the redesign of the street space were perhaps not completed.[18]

Thus, a little-maintained and little-used interim phase over perhaps even several decades seems identifiable between the two main

[16] Albrecht (2008: 114); Fabiani and Coccia (2003: 38); Döring-Williams and Albrecht (2020: 377–378), (2021: 170)

[17] Döring-Williams and Albrecht (2020: 377–380), (2021: 169–172)

[18] Döring-Williams and Albrecht (2020: 373, 376)

construction phases of Late Antiquity, i.e. between the Maxentian phase and the phase of the northern apse installation. However, also in this period the incompleteness of the building itself was related mainly to the stucco decorations and the outer covering of the façades—which can be excluded as a notable impact on the durability or vulnerability of the building.

3.3.2 Lack of Maintenance, Ageing, Natural Decay

Unlike wooden roof constructions, ageing has only little effect on Roman concrete (*opus caementicium*) structures, such as the walls and vaults of the Basilica of Maxentius. The solid vault constructions originally even had several waterproofing covering layers (see above).

3.3.3 Further Anthropogenic Interventions (Spoliation, Architectural Modifications, War, etc.)

The demolition of the shield wall in the central bay of the northern side aisle with the subsequent installation of the northern apse may be regarded as probably the most massive modifying intervention of the first, the Maxentian, construction. Alterations such as these, which significantly interfere with the original load-bearing structure, could of course have a strong impact on the vulnerability. However, because of all things, the northern side aisle with the northern apse did not collapse. This modification could obviously not have been causative.[19]

The only event of a war-like nature that can be considered hypothetically as the cause for major damages is the sack of Rome in 408–410 AD, an undoubtedly traumatic event.[20] But to date, there is no reliable evidence of destruction in the centre during the sacking.[21] And even if the Basilica of Maxentius had been affected, a collapse caused by plundering mercenaries can well be ruled out—unless there had been a major fire. But no such evidence could be detected in the whole complex.

3.3.4 Storm, Lightning, Intense Rain

Since considerable portions of the basilica's late antique masonry are still preserved *in situ* today—even without plaster and detached from the intact building context—and it only shows comparatively minor damage on the surfaces, the hypothesis of weather-related effects can be excluded as causative. Another argument against such an assumption is that rainwater systems played an important role in all construction phases of the building in Late Antiquity (see Döring-Williams and Albrecht 2020, Fig. 2). They have been integrated from the very beginning and have been verifiably considered in all structural alterations of this period.

3.3.5 Landslide, Flood, Hydrogeological Problems

In fact, hydrogeological problems are not to be expected in the location of the Basilica of Maxentius, originally situated in a relatively elevated position on Velian Hill.

3.3.6 Static Settlement

Static settlement is often incalculable and can lead to damage or even collapse if individual parts of the building settle to varying degrees. This is a most intricate phenomenon, which is discussed hypothetically below using the Basilica of Maxentius as a case study and distinguishing between four main causes.

[19] Amici (2005: 50–60) even claimed that the addition of the Northern apse was carried out at all to improve the statics (considered as 'abbastanza paradossale' by Coarelli 2010: 133)

[20] Lipps et al. (2013); Spera (2012)

[21] Lipps (2013)

3.3.6.1 Differing Soil Properties

In conjunction with the basilica's conservation and strengthening measures headed by the Soprintendenza alle Antichità di Roma for the *Grande Giubileo del 2000*, the subsoil and the basilica foundations were investigated quite intensively in the field of geotechnical engineering, presented, for instance by Calabresi in the 2005 monograph on the basilica (see also Sect. 3.3.6.3).[22] The aim was to understand the current static conditions and to assess the safety of the preserved northern side aisle with its three huge barrel vaults. Besides 12 geognostic boreholes in the soil, 13 cores were drilled vertically through the footing structures close to the principal piers to determine the depth of these structures.[23] Calabresi and Fattorini (2003: 452) summarised: 'The whole set of investigations allowed to define sufficiently accurate geotechnical sections, which showed the presence of the light brown silt layer under all the piers of the existing nave; this is followed by a layer of brown clay a few metres thick, then by a thick stratum of gravel and sand with clay intrusions; the Pliocene clay bedrock was found at about 35 m under the Basilica's plan'. The samples proved that while these different soil layers were more or less equally thick in the longitudinal direction, the thicknesses, however, in the transverse direction (especially at the locations of the six boreholes in the Via delle Carine along the western perimeter wall) differed remarkably by several metres and showed a slope angle of up to about 20° at the strata boundaries (see Calabresi 2005: Fig. 3.5c, p. 84).

Whether these layer differences can lead to differential settlement (due to the subsoil properties) would have to be evaluated by geotechnically experienced experts, but from an architect's point of view, this must not be completely ruled out. Furthermore, Mancini et al. (2014) showed that in deeper strata a buried valley axis (Middle Pleistocene) crosses the basilica in a diagonal direction.[24] This observation also leads one to suspect a rather unpredictable subsoil as a building ground.

3.3.6.2 Varied Loading

Differing loads arising from various structural elements can in principle lead to differential settlement. Accordingly, the more loaded area could settle more deeply than the less loaded area.

Within the basilica the loads from the superstructure inevitably vary as a result of the diverse types and dimensions of vaults used and very significantly, for example, between the small-span cross vaulted narthex and the wide-span barrel vaults in the side aisles.

Furthermore, the presence of diverse and very complex lateral thrusts from the vaults must be considered as an affecting factor. In the case of vaults in series—as, for example, in the case of the three barrel vaults of the side aisles—there is essentially no thrust on the transverse walls of the central bays, since the lateral load is equal in both directions. The situation of the outer transverse walls is quite different: massive and heterogeneous thrust exert here on the 'edges' of the building. In general, the foundations that transfer the loads from the superstructure to the ground have to be capable enough to avoid differential settlements so that the structure is safe and stable.

In total, Albrecht's[25] investigation of the basilica's foundations comprised 19 sections of varying extent and information content due to the respective circumstances. Seven of them were examined thanks to interim exposure during excavations that took place within the last two decades, and others thanks to the accessibility of sections exposed today. As a result, the principle structure of the basilica's foundations has been revealed. All foundations were designed as strip foundations (with individual widenings at the northwestern corner and along the western apse, which approximate a slab foundation design). The main loads of the basilica were supported by eight continuous north-south oriented footings. The foundations are made of coursed Roman concrete (*opus caementicium*) and

22 Calabresi (2005: 75–91)

23 Calabresi and Fattorini (2003); Calabresi (2005)

24 Mancini et al. (2014: Fig. 5)

25 Albrecht (2019)

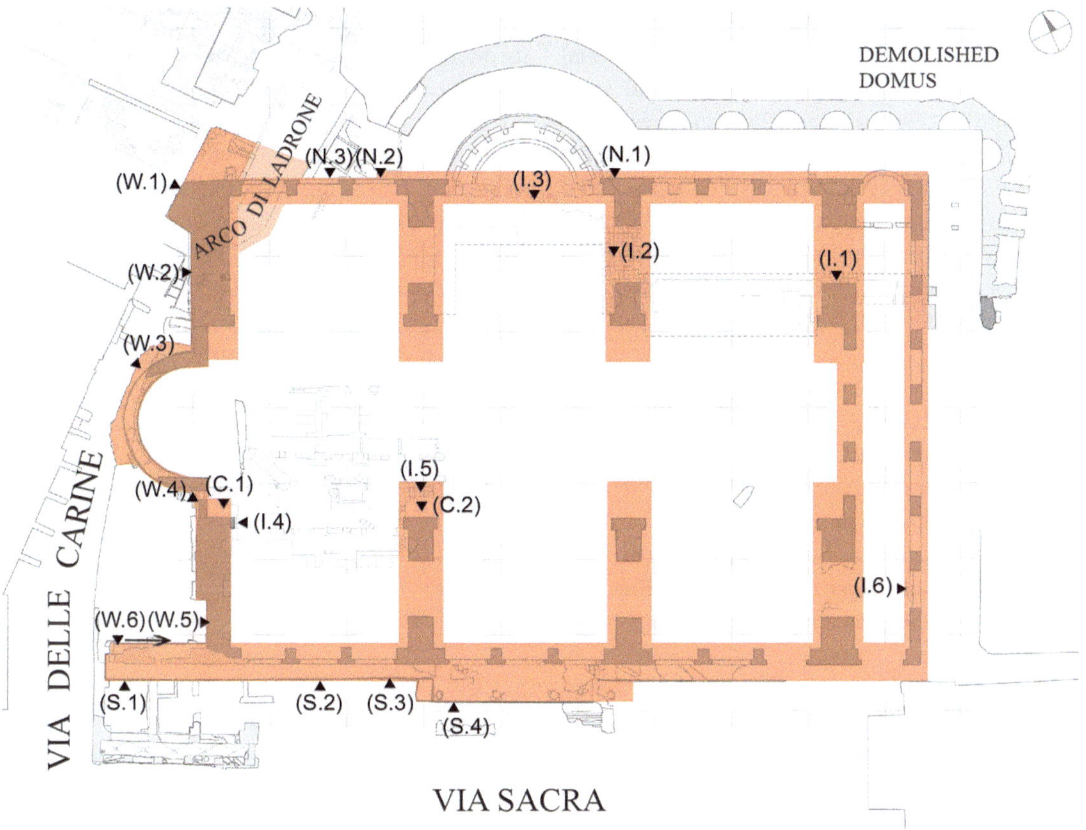

Fig. 3.5 Schematic reconstruction proposal of the basilica's foundation layout in plan; the remains of the *horrea piperataria* documented by Piranomonte (see Piranomonte 1996, 398) are considered in the west bays of the central nave and the south aisle; the investigated foundation sections are denominated (see Albrecht 2019)

typically exceed the footprints of the walls to the outside by between 0.6 m and 1.1 m. A schematic proposal of the foundation layout in plan is shown in Fig. 3.5.

The eight north-south oriented transverse walls of the aisles originally carried the major structure, thus the barrel vaults and major parts of the cross vaults of the central nave. The transverse walls, which are part of the western perimeter wall, are an exception. They are, according to the general topography, located at the edge between the ground level of the basilica and the street level of the Via delle Carinae (a difference in the height of about 7 m). So, even lateral earth pressure exerts the western perimeter walls. From a static perspective, however, the other six transverse walls consist of three zones each: two T-shaped huge piers and a large central arched opening in between.

The outer piers are structurally joined with the northern and the southern perimeter walls. The inner piers facing the nave not only carried the barrel vaults of the aisles, but also the impost blocks of the cross vaults of the central nave (see Fig. 3.2) and transmitted their loads to the strip foundations. Therefore, various loads were transmitted to the inner and outer portions of the transverse walls. Thus, differential settlements and corresponding effects here cannot be ruled out. Perhaps the crack that crosses the entire width of one of the transversal strip foundations, exactly in the axis of the arched opening, could be indicative of such a cause (shown by Samuelli Ferretti 2005a: 180, Fig. 6.21).

However, if the structural failure had been mainly the result of differential loading, an axial symmetrical collapse along the central axis should have occurred. Since the structures in the North are largely preserved, and those in the South have almost completely collapsed, the basilica's failure has, on the contrary, an extremely asymmetrical appearance; the hypothesis of a problem with differential loading can most likely be ruled out as the main cause of destruction.

3.3.6.3 Differing Compaction of the Subsoil or Substructures

It has been known for a long time that the detailed planning of the basilica was influenced considerably by the predecessor market structures beneath it—the *horrea piperataria.* According to Minoprio, as early as the 1930s this approach effected 'the maximum economy in foundations'.[26] The two main load-bearing Maxentian foundation sections in the *horrea piperataria* (Fig. 3.5, I4 and I5) could be investigated by Albrecht in 2004 and again in 2006.[27] In fact, the 'economy' of construction emphasised by Minoprio could be confirmed, but from a structural point of view, with frighteningly risky consequences. Thanks to modern consolidation substructures in reinforced concrete under the basilica, some areas of the formerly backfilled *horrea* are now accessible again. Here, despite the unavoidable adulterations and spatial restrictions associated with such extensive interventions, it is recognisable in several places, that both foundations were simply placed on top of the existing *pavimento in opus spicatum*, i.e. on top of the 'just left as is' brick floor of the predecessor structure laid in a herringbone pattern.[28] This floor originally was intended only to provide a clean and even walking surface for the market visitors of the former *horrea* with its rather low buildings and little open courtyards in between. A particularly compact subsoil was not necessary here. Therefore, the occurrence of a less compact material in this zone must be suspected, for example, in comparison to the natural ground or architectural structures in *opus caementicium*. Thus, from today's perspective, unpredictable settlements are quite conceivable in this zone of the structure. However, even the ancient builders could hardly have seriously expected a buried foundation underneath, or a ground that was particularly load-bearing. Thus, in this case, it is difficult to understand how such an obvious static weak point could have occurred or perhaps have been overlooked on the construction site of the Basilica of Maxentius, which otherwise presents itself as a project of superlatives.

3.3.6.4 Varied Foundation Structures and Dimensions

Another influencing factor when considering the respective settlements relates to the design of the individual foundations themselves. Normally, the design of the foundations should be developed to fit the requirements exactly and take into account both the specific loads and the subsoil. Examinations of the situation in the Basilica of Maxentius showed that the foundations here were each constructed in individual sections according to the progress of the entire construction.[29]

In the foundation sections she examined, Albrecht was able to collect data about the heights of the upper edges: in line with the general topography they vary between 18.74 m ASL at the western apse foundation (Fig. 3.5, W.3) and 26.60 m ASL at the narthex foundation (Fig. 3.5, I.6) at the east end.[30] However, the bottom edge could only be recorded in three cases (Fig. 3.5, S.4); (Fig. 3.5, I.4); (Fig. 3.5, I.5), respectively, a fourth bottom edge (Fig. 3.5, N.2) was likely to have been only slightly lower than that visible in the exposed part. The Maxentian foundation

[26] Minoprio (1932: 24)

[27] The documentation in 2004 was a side-product of Albrecht's postgraduate thesis on the remains of the collapsed vaults, i.e. vault fragment I in the *horrea piperataria*. Access was granted again in 2006 to conduct a masonry analysis of the Maxentian foundations.

[28] Albrecht (2019: 182–183)

[29] Albrecht (2019: 185)

[30] Albrecht (2019: 184)

depth amounts to 2 m, 2.40 m, 2.60 m, and slightly more than 2.90 m. Of those, only the latter value arises from a foundation structure laid in natural ground, whereas the other three foundations are laid upon previous structures.

In summary, the few but reliable measurement results of the Maxentian foundation's depths only range from 2 m to slightly more than 2.90 m, documented in different areas and load cases, both at the northern and southern perimeter walls and at the strip foundation separating the bays of the side aisles.

Calabresi and Fattorini, based on the examination of core drillings, interpreted the situation as follows: 'Every pier has its own foundation; their depth increases from east to west according to the Velia's natural slope, from 4 m below the Basilica's plan (*Colosseum* pier) to 14 m (*Carinae* pier)'.[31] This striking discrepancy between the measurements is probably due to the fact that the assessment of Calabresi's drill cores apparently did not (clearly) distinguish between material from Maxentian foundations, heterogeneous building fabric of predecessor structures and backfilling.[32] The foundation's depths here, therefore, appear generalised and with an unexplained response to the slope of the terrain. The values then were obviously extended and visually presented as a reconstruction proposal[33] for the entire foundation design,[34] which in turn has been used as the basis for safety evaluations even to date[35] or perhaps even for the current construction sites of the 'Metro linea C'.

While Calabresi relates the values to the floor level of the basilica, Albrecht considers the values from the structural upper edge of the foundation.[36] Nevertheless, despite different parameters, approaches, and evaluation methodologies, both investigations demonstrate that identical dimensioning was not maintained for features subjected to the same load cases. Thus, the foundations themselves could also have contributed to differential settlement, but the asymmetrical damage pattern again evidently indicates that this hypothesis was not decisive.

3.3.7 Building Material and Construction Related Failures, Miscalculations

Preliminary note: A collapse due to structural reasons other than differential settlement (without additional load effects) would be feasible during the construction work or at the latest until the mortar/*opus caementicium* has set. The duration of the setting process in the ancient *opus caementicium* is mainly dependent on the pozzolanic activity. However, as this pozzolanic reaction must be considered as a long-term one, at least some possible causes of material-related failure with respect to *opus caementicium* are discussed here.

The nave's western perimeter walls with the western apse were apparently the structurally weakest zone of the complex. One has to be aware that the late antique Via Sacra, running more than 100 m along the basilica, was quite steeply inclined, ranging from 19 m above sea level (ASL) in the west at the junction of Via delle Carine into the Via Sacra, up to 26.7 m ASL in the east in front of the Temple of Venus and Roma (Fig. 3.6). Since the design concept for the basilica included an entrance from the east, the new floorplan was laid almost at this height of 26.70 m ASL. The height difference increasing towards the west was then compensated by massive, metre-high earth backfilling, forming a kind of

[31] Calabresi and Fattorini (2003: 449). The authors strongly advise to doubt the value (14 m) at the Carinae pier, since a reliable value is documented only up to 9 m below the basilica's floor level. Deeper structures probably originate from cores that penetrated predecessor structures.

[32] See, for instance, drilling point F12 in Calabresi and Fattorini (2003: 452) and the critics on that misleading value in Albrecht (2019: 175)

[33] On a general methodological comment about the reconstruction proposal, see Albrecht (2019: 175)

[34] Calabresi (2005: 79)

[35] E.g. Albuerne (2016: Fig. 2.8, p. 43; 62); Calabresi (2005: 86–91); Pau and Vestroni (2013: 456); Samuelli Ferretti (2005a: 194; 200–225); Samuelli Ferretti (2005b: 233–257)

[36] Additionally the values are given in metres above sea level (ASL).

Fig. 3.6 Sloping course of the late antique Via Sacra and junction of Via delle Carine into it; the photo shows the pre-excavation situation in the last quarter of the nineteenth century. Photo: AERIA (Antikensammlung Erlangen Internet Archive): https://web.archive.org/web/20101210092926/ http://www.aeria.phil.uni-erlangen.de/photo_html/topographie/italien/rom/forum_romanum/forro19.html (Consulted on 27th March 2023)

levelling podium. The backfilling reaching its highest dimension along the Via delle Carine was retained by the 7 m high substructures of the western perimeter walls, which even integrated sections of the predecessor market buildings. Here, of course, the lateral load of the backfilling has a certain effect on these substructure walls (see Sect. 3.3.6.3).

In the search for possible structural deficiencies, the wall thicknesses here play a further significant role. The wall thickness on the western wall of the nave is only 6 Roman feet at the substructure level, and as little as 5 ft above that (compared to ~16 ft in the northern aisle area). So, the thickness of the western nave walls roughly corresponds to that of the northern and southern longitudinal perimeter walls. The latter, however, as shield walls of the barrel vaults spanning them—in contrast to the western and eastern nave walls—only had to bear the dead load. The western (and eastern) nave walls, though, had to support considerable partial loads resulting from the thrust of the cross vaults—cross vaults, whose ridge height was enormous: originally about 40 m above the floor level of the basilica.[37]

As mentioned above, Albuerne and Williams applied graphic statics to assess the structural behaviour of the original vaults, in particular that of the nave's cross vault.[38] 'A thrust line analysis of the reconstructed cross vault using a conservative slicing approach has confirmed that

[37] Albrecht (2004: 86)

[38] Albuerne (2016: 53–55); Albuerne and Williams (2017: 910–911)

Fig. 3.7 (**a**, **b**) Vault Fragment VII next to the wall of the *Forma Urbis Romae*. Vault Fragment VII originally belonged to the west apse calotte of the Basilica of Maxentius. Photos: L. Albrecht

the vault was stable under gravity loads', they concluded their investigations.[39] By using stability analysis, they approached the issue of the basilica's load-bearing capacity from an engineering perspective. But, generalisation and simplification during virtual 3D reconstruction can generally not be avoided. For example, the understanding of load distribution within vaults, which is inherently difficult already in 'normal' *opus caementicium* conglomerate, is made particularly difficult by the fact that in addition to the coursed *opus caementicium*, radial brick ribs, the so-called *nervature*, have been inserted. Since some vault fragments were observed to have collapsed along such *nervature* edges,[40] their influence on the specific transfer of loads may be currently underestimated.

In the following, some selected structural details will be pointed out that would be worth including for later remodelling.

In 2004, Albrecht succeeded in identifying a hitherto neglected vault fragment (3.60 m × 0.90 m × 1.60 m), Fragment VII, now located next to the wall of the *Forma Urbis Romae* in the aula of the Templum Pacis (Fig. 3.2: VII; Fig. 3.7), as originally belonging to the west apse calotte of the Basilica of Maxentius.[41] The upper fracture edge runs along a *bipedales* course (Roman bricks measuring two Roman feet per side), which ran originally as a horizontal course through the entire vault thickness. In general, in Late Antiquity two different construction methods for *caementicium* domes are found: those containing a dense *nervature* rib system ('Kammersystem')[42] and those primarily using bonding courses in combination with a series of *nervature* ('Bündelsystem')[43] only at main load-bearing positions. The *bipedales* course can be used as evidence for the application of the latter method for the basilica's apse vault.

Constructively the most remarkable aspect here is the special material selection of the *caementa* (the rubble) in the *opus caementicium*:

[39] Albuerne and Williams (2017: 911)

[40] Shown for Fragment I and Fragment III in Albrecht (2008: 111). Fragments II, IV, and V also show *nervature* at certain edges.

[41] Fragment VII in Albrecht (2004: 72–81, 108–115), (2008: 115–116); compare also Döring-Williams and Albrecht (2021: 163)

[42] Rasch (1991: 373–378)

[43] Rasch (1991: 372–373)

Fig. 3.8 Basilica of Maxentius, western nave wall above the level of the apse's vault springers; the *opus caementicium* shows a large amount of heavy leucititic lava *caementa*. Photo: E. Stamm, L. Albrecht

the *caementa* here consists of about 50% of leucititic lava rubble, about 30% of brick and a small amount of tuff, travertine, and marble. The 'leucititic lava (2800 kg/m^3), commonly called *selce*, (...) is extremely hard and difficult to quarry and was used primarily for road building or as *caementa* in foundation walls'.[44] Thus, the unit weight of the lava is about twice as heavy as the tuff.

It is remarkable that the same lava rubble had been observed at the western nave wall, where a change to significantly heavier material exists exactly at the level of the vault springers of the western apse (Fig. 3.8).[45] The lower wall core here shows the 'normal' mixture made of tuff, bricks, some travertine, and some other materials as in most of the Maxentian wall zones. Obviously, the material weight was to be increased here. In fact, it was general practice to grade the *caementa,* but usually the heavier ones were used in foundations and the lighter ones in vaults.[46] Here, on the contrary, it is the other way around. Lancaster rightly concludes that the 'use of the heavy leucititic lava was perhaps an attempt to provide a surcharge to counter the thrusts of the vaults'.[47]

For the area of the western nave wall, however, a number of influencing variables remain indeterminable, since most of the area is lost. These include, for example, the question of how the extrados shape of the apse calotte looked like, the question of whether there was a weight reduction at the apsis crown—as was usually the case[48]—and the question of whether a window existed above the apse.

In summary, it becomes clear how structurally borderline this wall had been built, but also that the builders were apparently (partly) aware of this problem. They reacted to this problem at least to

[44] Lancaster (2005: 16)

[45] See also Lancaster (2005: 198)

[46] Lancaster (2005: 59)

[47] Lancaster (2005: 198)

[48] Lancaster (2005: 213–214)

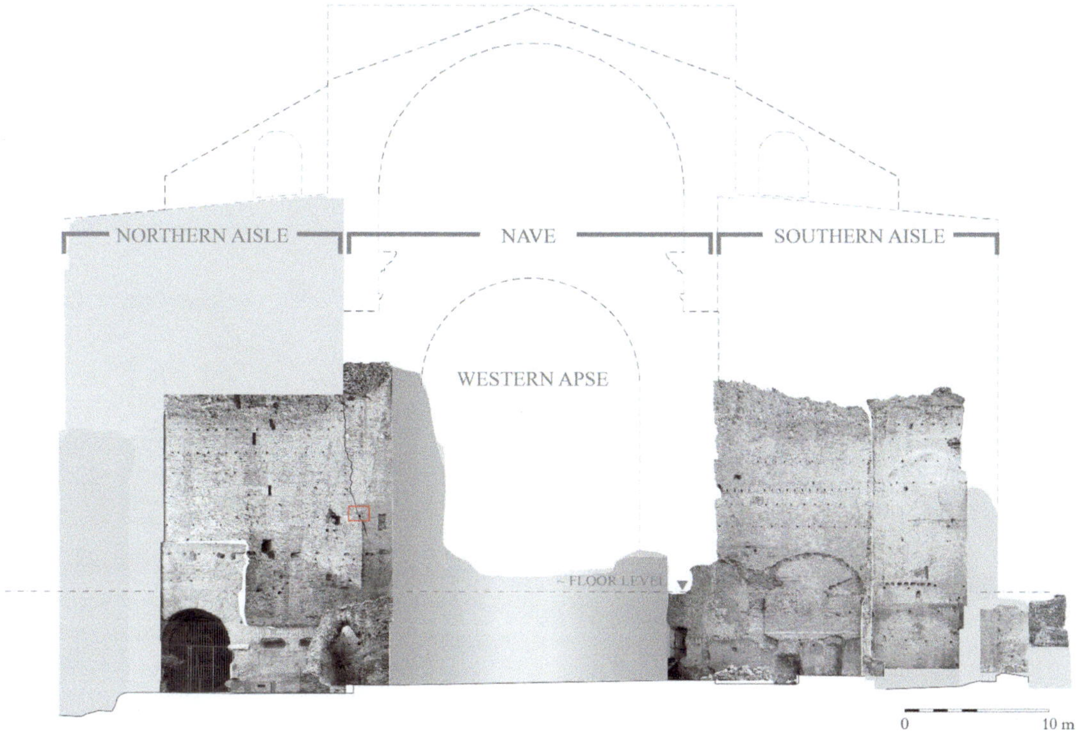

Fig. 3.9 Basilica of Maxentius, Rome, western perimeter wall along the Via delle Carine. Photogrammetric rectification of the walls (fokus GmbH Leipzig) of the northern and southern aisles, sections outside the rectification level shaded grey, view from west. The red frame marks the area of Fig. 3.10. Dashed: hypothetical contour lines of the basilica's nave vault, roof, and western apse added by the authors

some extent and made improvements during construction. It seems quite obvious that a potential trigger that could have caused this structurally questionable, fragile wall to collapse did not have to be particularly severe. Again, from today's perspective, one wonders about the discrepancy between such obviously risky, even near-negligent design decisions on the one hand, and the knowledge of material properties and statics that is evident in many cases on the other.

3.3.8 Seismic Events

Certainly, the Basilica of Maxentius shows cracks. The most prominent one is located in the direct extension of the northern nave wall at the corner with the western nave wall (Figs. 3.9 and 3.10). Furthermore, the corresponding western wall shows out-of-plumb displacements towards the west.[49] Distinctive 'typical' effects of earthquakes such as rotated and displaced buttress walls tilted, displaced, or folded walls could not be observed so far in the preserved Maxentian structure, and special fracture patterns such as penetrating fractures in wall blocks, horizontally running or conjugate fractures could only be vaguely determined.[50] The significance of a sudden seismic effect here is vague not least because the process of deformation could also have

[49] According to Albuerne, the displacement measures about 0.87°, Albuerne (2016: 43). The leaning documented in our own measurements is even slightly larger (~1°).

[50] For a general depiction of earthquake effects, see Fig. 2 in Rodríguez-Pascua et al. (2011: 22).

Fig. 3.10 Basilica of Maxentius, crack in the western perimeter wall in direct extension of the northern nave wall. Photo: L. Albrecht, G. Eßer, I. Piironen

occurred or continued, delayed or complete, after the main collapse. For example, the thrust from the barrel vault due to a loss of substance was no longer countered by parts of the original superimposed load and may have led—possibly with a time lag and as a domino effect—to further deformations of the building fabric.

In an attempt to provide evidence that a seismic event may have triggered the collapse of the basilica, Albuerne has described supposed earthquake damage characteristics in the truncated pillars of the *horrea piperataria* and thus only within the predecessor structures.[51] Based on the damage characteristics observed specifically here, she was able to plausibly determine that the destructive lateral (earthquake-typical) loadings occurred at a time when the *horrea* spaces were no longer intact.

In concrete terms, this observation merely means that the destruction of the *horrea piperataria* represents the *terminus post quem* the site was exposed to seismic loading. However, a simultaneity with the partial collapse of the basilica itself cannot yet be proven to date and must not be taken for granted.

Another opportunity to confirm or exclude an earthquake as causative could be derived from the original positions of the collapsed structures. The difficulty here, however, is the fact that the collapse occurred a very long time ago and that the site has undergone many changes in the meantime, such as cleanings, spoliation, and numerous excavation campaigns. At the latest in the

[51] Albuerne (2016: 63–71)

Renaissance, excavations have penetrated at least to the level of the lost statue pedestal in the western apse.[52] Still, the exact position, even of the individual parts of the statue itself has not survived with certainty. It was only in 1811–1813 that the central nave, the side aisles, and parts of the narthex were cleared (again) of the rubble, which at that time was still up to 25 ft high.[53] During this campaign, P. A. Pâris and M.-P. Gauthier were able to survey and to document some significant fragments of the collapsed nave's cross vaults.[54] Both compiled corresponding reconstruction proposals, but neither of them documented the collapsed positions. Up to today, 25 distinct fragments have survived within the basilica or in its immediate vicinity.[55] The majority of the fragments do not have a characteristic shape, but seven fragments clearly show parts of a vault intrados or a roof surface and could thus be identified as original vault or roof elements. They were documented and analysed in detail. Their present position is shown in the plan (Fig. 3.2). Of these seven fragments, two presumably remain 'untouched' and *in situ* in the collapsed position (Fragment II and Fragment V), two were certainly moved (Fragment III and Fragment IV). Fragments I and VI are in their original position but were underlaid with masonry after the excavation activities were completed.

To our knowledge, the mentioned Fragment VII (see Sect. 3.3.7. and Fig. 3.7) has not yet been examined about characteristics of potential layers at its bottom. Nevertheless, the fragment could be identified as part of the calotte of the western apse.[56] The linear distance between the original position in the vault and the present position is quite large, about 17 m (taken in the floor plan) (see Figs. 3.2 and 3.7). An excavation in the area in question was carried out by Efisio Luigi Tocco in 1867. This was more of a search 'dig' and clearly made in haste since the only goal was the discovery of new fragments of the *Forma Urbis Romae*.[57] These facts strongly suggest that Fragment VII is no longer in its original collapsed position. Nevertheless, it is remarkable in several respects: If the building had 'simply' collapsed, the vault area of the western apse would have had to fall inward this apse. But in this case, the fragment lies more to the west and outside the building. Particularly noteworthy, however, is one of the fracture edges, which contains remnants of a *bipedales* (bonding) course (see Fig. 3.7).[58] *Bipedales* courses usually were laid horizontally and ran through the entire wall (or vault) thickness, e.g. as regular levelling courses. This observation on the horizontal course of the fracture of Fragment VII therefore can be interpreted as an indication of the impact of lateral loads—and thus, perhaps, of a seismic cause.

3.4 Conclusion

In this paper, possible triggers for the collapse of the Basilica of Maxentius were systematically discussed. All the aspects described can stress or damage a building fabric. Here, they were examined for their possible relevance to this selected building (see Table 3.1). The previous results of investigations from a wider range of engineering sciences were also assessed. The focus of this work was on static settlements and on selected structural phenomena observed at the basilica itself. In summary, the following results can be noted:

The interpretation that the site was affected by a seismic event could most likely be confirmed, even if very few legible traces are preserved on the building remains and in the surrounding area. The position and characteristic fracture edge of the discussed Fragment VII of the west apse's calotte eventually could be counted here as an example.

[52] Buddensieg (1962)

[53] Uginet (1985: 212); Pinon (2007: 107)

[54] See Fig. 6 in Döring-Williams and Albrecht (2021).

[55] In her master's thesis, Albrecht identified 25 fragments as originally belonging to the Maxentian basilica.

[56] Albrecht (2008: 116–117)

[57] Tucci (2018: 308–309)

[58] Albrecht (2008: 115)

Table 3.1 Evaluation of potential responsibility for damage of the Basilica of Maxentius based on building archaeological assessment: Without evidence, –, Not completely ruled out (x), Some evidence and perhaps co-responsibility x; causes proposed as most likely, xx

3.1	Unfinished structure/roofing		–
3.2	Lack of maintenance, ageing, natural decay		–
3.3	Further anthropogenic interventions (spoliation, architectural modifications, war, etc.)		–
3.4	Storm, lightning, intense rain		–
3.5	Landslide, flood, hydrogeological problems		–
3.6	Static settlement	1. Differing soil properties	x
		2. Varied loading	(x)
		3. Differing compaction of the subsoil or substructures	xx
		4. Varied foundation structures and dimensions	(x)
3.7	Building material and construction related failures, miscalculations		xx
3.8	Seismic events		xx

Anyway, the archaeological and building investigations in the east of the complex have provided evidence that parts of the narthex, and as a consequence possibly further adjoining/statically interdependent parts of the building (presumably including the nave's vaults), were destroyed as early as the fifth century. Whether the damages found at the narthex and in the *horrea piperataria* result from one single static failure can be assumed but has not yet been proven.

The investigations have also shown how alarmingly weak the west walls were in terms of wall thickness and foundation, but with a total height of up to 47 m above street level. In particular here, where a partial thrust of the cross vaults exerted the walls, the statics of the substructures were additionally weakened by lateral earth pressure, which resulted from the backfilling of the *horrea piperataria.*

At the foundation level of the basilica, this structurally precarious situation was further exacerbated by the apparently difficult-to-calculate subsoil since at least some of the basilica's foundations within the former market place were apparently simply placed on top of the predecessor pavements or in between the demolished building fabrics without digging down to clearly solid ground.

These factors suggest that the Basilica of Maxentius, which was planned as a prestige project for the emperor and which Aurelius Victor had still described as 'magnifice construxerat' around the middle of the fourth century,[59] must have been a kind of timebomb in terms of structural weakness. Whether an earthquake could eventually have been responsible for the collapse, or whether 'only' a slow settlement could have been sufficient to cause the damage, cannot be answered with certainty from a building archaeological perspective. The described horizontal fracture edge (Fragment VII) may suggest that the shear forces here were induced by a sudden impact, as in an earthquake, rather than during decades of gradual exposure. However, most likely, the fragile structure was already so weakened by a combination of settlement causes (in particular differing soil properties and different compaction of the subsoil or substructures in the southern side aisle); also, a minor trigger of a seismic movement could have been sufficient to cause a large part of the structure to collapse. Against this background, a relatively early collapse of the Basilica of Maxentius, which is now proposed to have occurred as early as the fifth century, does not seem implausible.

[59] Aur. Vict. Caes. 40, 26: *Adhuc cuncta opera, quae magnifice construxerat, urbis fanum atque basilicam, Flavii meritis patres sacravere*; cited in Ziemssen 2011, 25.

References

Albrecht L (2004) Italien, Rom: Untersuchung der Gewölbe- und Dachfragmente der Maxentiusbasilika. Unpublished master's thesis. Technische Universität Berlin

Albrecht L (2008) Die Gewölbe der Maxentiusbasilika. Neue Ergebnisse aus den Fragmentuntersuchungen. In: Koldewey-Gesellschaft (ed) Bericht über die 44. Tagung für Ausgrabungswissenschaft und Bauforschung. Vom 24. bis 28. Mai 2006 in Breslau. Rudolf Habelt, Bonn, pp 107–117

Albrecht L (2019) The foundations of the basilica of Maxentius in Rome: a study of late antique building practices. In: Campbell J et al (eds) Water, doors and buildings: studies in the history of construction. The proceedings of the sixth conference of the Construction History Society. Lulu, Cambridge, pp 173–188

Albrecht L, Döring-Williams M (2010) Die Maxentiusbasilika als Ruine in Spätantike und Mittelalter. In: Perlich B, van Tussenbroek G (eds) Mittelalterliche Architektur. Bau und Umbau, Reparatur und Transformation; Festschrift für Johannes Cramer zum 60. Geburtstag. Imhof, Petersberg, pp 195–218

Albuerne A (2010) The stability of the basilica of Maxentius in Rome. In: Gu X, Song X (eds) Selected, peer reviewed papers from the 7th international conference on structural analysis of historic constructions, October 6–8, 2010, Shanghai, People's Republic of China. Structural analysis of historic constructions; strengthening and retrofitting; Part 1, pp 325–330. Trans Tech Publications, Stafa-Zurich

Albuerne A (2016) Seismic collapse of vaulted structures: unreinforced quasi-brittle materials and the case study of the basilica of Maxentius in Rome. PhD dissertation, University of Oxford

Albuerne A, Williams MS (2017) Structural appraisal of a Roman concrete vaulted monument: the Basilica of Maxentius. Int J Arch Herit 11(7):901–912

Amici CM (2005) From project to monument. In: Giavarini C (ed) The Basilica of Maxentius. The monument, its materials, construction, and stability (Studia archaeologica 140). L'Erma di Bretschneider, Rome, pp 21–74

Barosso M (1940) Le costruzioni sottostanti la Basilica Massenziana e la Velia. In Atti del 5. Congresso nazionale di studi romani, 58–62; pl. XII–XV. Istituto di studi romani, Rome

Buddensieg T (1962) Die Konstantinsbasilika in einer Zeichnung Francescos di Giorgio und der Marmorkoloss Konstantins des Grossen. Münchener Jahrbuch der bildenden Kunst 13:37–48

Buzzetti C, Virgili P (1985) Notiziario di scavi e scoperte: Regione IV. Velia. Planimetrie dal Guglielmo Gatti (1932). Bullettino della Commissione Archeologica Comunale di Roma 90(2):314–320

Calabresi G (2005) Subsoil and foundations. In: Giavarini C (ed) The basilica of Maxentius. The monument, its materials, construction, and stability (Studia archaeologica140). L'Erma di Bretschneider, Rome, pp 75–91

Calabresi G, Fattorini M (2003) Safety assessment of the foundations of the basilica of Maxentius in Rome. In: Brebbia CA (ed) Structural studies, repairs and maintenance of heritage architecture (WIT Transactions on the Built Environment 66). WIT Press, Southampton, pp 447–460

Coarelli F (2010) La basilica di Massenzio e la praefectura urbis. In: Bonamente G (ed) Istituzioni, carismi ed esercizio del potere (IV-VI secolo d.C.). Edipuglia, Bari, pp 133–146

Döring-Williams M, Albrecht L (2020) Die Nordapsis der Maxentiusbasilika. Eine Neuinterpretation der Baubefunde. In: Piesker K, Wulf-Rheidt U (eds) Umgebaut. Umbau-, Umnutzungs- und Umwertungsprozesse in der antiken Architektur. Internationales Kolloquium in Berlin vom 21.–24. Februar 2018 veranstaltet vom Architekturreferat des DAI (Diskussionen zur archäologischen Bauforschung 13). Schnell & Steiner, Regensburg, pp 365–382

Döring-Williams M, Albrecht L (2021) *Magnifice construxerat.* Die Innenausstattung der Maxentiusbasilika in Rom. In: Ruppienė V (ed) Stone and splendor: interior decorations in late-antique palaces and villas. Proceedings of a workshop, Trier, 25–26 April 2019 (Forschung zu spätantiken Residenzen 1). Harrassowitz, Trier, pp 155–175

Du Pérac E (ed) (1621) I vestigi dell'Antichità di Roma: Raccolti et ritratti in perspettiva con ogni diligentia. Scaichi, Roma

Fabiani AG, Coccia S (2003) Le indagini archeologiche recenti, in Centro Interdipartimentale di Scienza e Tecnica per la Conservazione del Patrimonio Storico-Architettonico. In: Atti del convegno su: La Basilica di Massenzio. Ricerca interdisciplinare applicata allo studio e alla conservazione di un monumento. CISTeC, Rome, pp 30–51

Galadini F, Hinzen KG, Stiros S (2006) Archaeoseismology: methodological issues and procedure. J Seismol 10(4):395–414

Galadini F, Ricci G, Falcucci E, Panzieri C (2018) Archaeoseismological evidence of past earthquakes in Rome (fifth to ninth century A.D.) used to quantify dating uncertainties and coseismic damage. Nat Hazards 94(1):319–348

Galli PAC, Molin D (2014) Beyond the damage threshold: the historic earthquakes of Rome. Bull Earthq Eng 12(3):1277–1306

Giavarini C (ed) (2005) The Basilica of Maxentius. The monument, its materials, construction, and stability (Studia Archaeologica 140). L'Erma di Bretschneider, Rome

Lancaster LC (2005) Concrete vaulted construction in imperial Rome: innovations in context. Cambridge University Press, Cambridge

Lipps J (2013) Alarichs Goten auf dem Forum Romanum? Überlegungen zu Gestalt, Chronologie und Verständnis der spätantiken Platzanlage. In: Lipps J, Machado C, von Rummel P (eds) The sack of Rome in 410 AD: the event, its context and its impact. Proceedings of the conference held at the German Archaeological Institute at Rome, 04–06 November 2010. Reichert, Wiesbaden, pp 103–122

Lipps J, Machado C, von Rummel P (eds) (2013) The sack of Rome in 410 AD: the event, its context and its impact. Proceedings of the conference held at the German archaeological Institute at Rome, 04–06 November 2010. Reichert, Wiesbaden

Mancini M et al (2014) A physical stratigraphy model for seismic microzonation of the central archaeological area of Rome (Italy). Bull Earthq Eng 12(3): 1339–1363

Minoprio A (1932) A restoration of the Basilica of Constantine, Rome. Papers of the British School at Rome 12:1–25

Pau A, Vestroni F (2013) Vibration assessment and structural monitoring of the Basilica of Maxentius in Rome. Mech Syst Signal Process 41(1–2):454–466

Pinon P (2007) Pierre-Adrien Pâris (1745–1819), architecte, et les monuments antiques de Rome et de la Campanie. École Française de Rome, Rome

Piranomonte M (1996) Horrea Piperataria. In: Steinby EM (ed) Lexicon Topographicum Urbis Romae, vol 3, pp 45–46. Fig 32, p 398

Rasch JJ (1991) Zur Konstruktion spätantiker Kuppeln vom 3. bis 6. Jahrhundert. Jahrbuch des Deutschen Archäologischen Instituts 106:311–383

Rodríguez-Pascua MA et al (2011) A comprehensive classification of earthquake archaeological effects (EAE) in archaeoseismology: application to ancient remains of Roman and Mesoamerican cultures. Quat Int 242(1): 20–30

Samuelli Ferretti A (2005a) The structures of the basilica. In: Giavarini C (ed) The Basilica of Maxentius. The monument, its materials, construction, and stability (Studia archaeologica 140). L'Erma di Bretschneider, Rome, pp 161–225

Samuelli Ferretti A (2005b) Proposals for a permanent strengthening. In: Giavarini C (ed) The Basilica of Maxentius. The monument, its materials, construction, and stability (Studia Archaeologica 140). L'Erma di Bretschneider, Rome, pp 227–257

Spera L (2012) La realtà archeologica: restauro degli edifici pubblici e riassetto urbano dopo il sacco, in Roma e il sacco del 410: realtà, interpretazione, mito. Atti della Giornata di studio (Roma, 6 dicembre 2010). (Studia ephemeridis Augustinianum), pp 113–155. Istituto Patristico Augustinianum, Rome

Tucci PL (2018) 5.3 The excavation in the hall of the Forma Urbis (July 29–September 28, 1867). In: Tucci PL (ed) Efisio Luigi Tocco. A 'lover of antiquities' in 19th-century Rome. Edizioni Quasar, Rome, pp 297–332

Uginet FC (ed) (1985) Roma antiqua: envois des architectes français (1788–1924); Forum, Colisée, Palatin; Curie (forum romain), Villa Médicis, Rome, 29 mars–27 mai 1985, Ecole Nationale, Supérieure des Beaux-Arts, Paris, 7 mai–13 juillet 1986. Académie de France à Rome

Ziemssen H (2011) Das Rom des Maxentius. Städtebau und Herrscherbild zu Beginn des 4. Jh. n. Chr. PhD dissertation, Universität Hamburg

Part II

Identification and Classification Methods

4 The Archaeology of Post-Earthquake Repairs: New Insights into Building Processes in Roman Pompeii

Hélène Dessales, Julien Cavero, Agnès Tricoche, and Laura Pecchioli

Abstract

Repairs are important evidence for analysing the impacts of an earthquake in a built environment, and their study is one of the main issues of archaeoseismology. This paper proposes a systematic analysis of post-seismic repairs through the fertile testing ground represented by the Roman city of Pompeii. It sets out the methodology of the research programme that was carried out (ANR RECAP, 2015–2019), which involved three main steps: the typology of architectural repairs and reconstructions; the analytic criteria of the repairs and building techniques through the specific database OPUR (*Outils pour unités de réparation*; Tools for repair Units); and the cartography of post-earthquake interventions in Pompeii. This protocol could be extended to other ancient sites or historic buildings in prone-earthquake areas, regardless of the period considered.

Keywords

Reconstruction · Repair · Building techniques · Earthquake · Risk · Database · GIS · Pompeii

H. Dessales (✉)
École normale supérieure, AOROC (Archéologie et Philologie d'Orient et d'Occident), Université PSL (Paris Sciences & Lettres), Paris, France
e-mail: helene.dessales@ens.psl.eu

J. Cavero
CNRS (Centre national de la recherche scientifique), MOM (Maison de l'Orient et de la Méditerranée), Lyon, France

A. Tricoche
CNRS (Centre national de la recherche scientifique), AOROC (Archéologie et Philologie d'Orient et d'Occident), Université PSL, Paris, France

L. Pecchioli
Institut für Archäologie, Klassische Archäologie, Winckelmann Institut, Ostia Forum Projekt (OFP), Humboldt Universität zu Berlin, Berlin, Germany

Institut für Kunstgeschichte, Bauforschung und Denkmalpflege, Technische Universität Wien, Vienna, Austria

Dipartimento di Architettura e Restauro, Studi Urbani di Roma, Università di ROMA TRE, Rome, Italy

4.1 Introduction

A repair aims to restore the functionality of a building or part of a building that has been damaged through wear, a defect, or an accident. In his treatise *De re aedificatoria* (1452), Alberti devotes an entire section to the repair of buildings (Book X), in which he specifies various techniques to ensure the durability of a monument, facing the effects of erosion or earthquakes or original architectural design errors.[1] A repair requires deterioration as a necessary condition for its implementation and implies a distinction with maintenance, which aims, on the contrary, to

[1] Alberti (1988).

L. Pecchioli (ed.), *Archaeoseismology*, Natural Science in Archaeology,
https://doi.org/10.1007/978-3-031-28303-1_4

maintain the object in the same state by anticipating deterioration.[2] While there is no fundamental difference between repairing and maintaining, both acts present a different temporality about the works and the damage: repair occurs after damage, whereas maintenance aims to avoid it. In other words, the first one is curative; the second one is preventive. In Roman Antiquity, particularly in legal texts, a similar distinction is made between the *refectio*, the repair of damage, and the *tutela*, defined as the prevention of damage, daily maintenance, and minor repairs. Both interventions guarantee the good conservation of a building, defined by the concept of *sarta tecta*.[3]

Repairs are important evidence for analysing the impacts of an earthquake and the economy of reconstruction. Indeed, their study is one of the main issues of archaeoseismology.[4] In the comprehensive classification of earthquake archaeological effects (EAEs), they are defined as secondary or indirect seismic effects, whereas destruction layers or structural damage are considered primary effects.[5] In his study of the specific Minoan context, Simon Jusseret adapted the EAE classification, integrating the criteria of potential earthquake archaeological effects (PEAEs), and defined repairs as structural post-seismic PEAEs: patched walls, block doorways, recycling construction materials, anti-seismic constructions.[6]

This paper proposes a systematic approach for post-earthquake repairs through the fertile testing ground represented by the Roman city of Pompeii, which provides an exceptional opportunity to examine building processes. The typology of repairs will be defined, and two main complementary tools will be presented: (1) a specific database for analysing the repairs and (2) the cartography of the post-seismic interventions in Pompeii. Combining both tools provides insight into the economics of post-seismic reconstruction in Roman times.

4.2 The Typology of Repairs

In the case of post-seismic repairs, the solutions vary greatly, depending on the context, from emergency patching up to well-articulated large-scale projects. They depend not only on economic and cultural conditions but also on the frequency of seismic incidents, which could generate a major sense of vulnerability in a specific place, and lead to technical innovations.[7] Three significant factors determine this culture of repair: the level of qualification of the builders[8]; the availability of manpower and materials; and the scope to finance the building processes.[9] The RECAP programme (2015–2019), funded by the French Agence Nationale de la Recherche, addressed these key questions by tracing Roman post-earthquake reconstructions.[10] To develop this approach, the project focused on the emblematic site of Pompeii. Indeed, this small town of Campania suffered several earthquakes which followed one another at short intervals in the 20 years before the eruption of Vesuvius in 79 CE. The first, dated 62/63 CE and probably the most violent,[11] is well-known thanks to ancient written sources, whereas the subsequent episodes have only been identified through archaeological data.[12] Such a case study demonstrates the “resilience” of a Roman town,[13] and all the process of urban reconstruction, brutally interrupted by the final eruption. Focusing on the repairs carried out after an earthquake allows us to explore such a building system implemented in times of crisis and emergency.

2 L'Héritier et al. (2018).

3 Davoine et al. (2019).

4 Jusseret and Sintubin (2017).

5 Rodriguez Pascua et al. (2011).

6 Jusseret et al. (2013); Sintubin (2017, 83–84).

7 Ferrigni et al. (2005).

8 Dessales (2012).

9 Dessales (2016).

10 ANR-14-CE31-0005 (2015–2019). Cf. Recap (2005) and Dessales (2022).

11 Savino (2009); Guidoboni and Ebel (2009, 425–428).

12 Fröhlich and Jacobelli (1995); Varone (2005); Dessales (2020a, 427–446).

13 Monteix (2011).

The case of Pompeii is especially suitable for investigation because of the extent of the remains (64 ha within the city, 3/5 of which have been excavated) and the quality of conservation of the buildings and their decoration. In this context, a small scale can be considered for specific construction sites in different types of buildings, instead on a large scale provides a complete overview of the management of building processes in the ancient town.

On a small scale, the RECAP programme explored two complementary case studies: firstly, priority urban services evidenced by the water network as well as through the extra-urban aqueduct and the fifteen public urban water towers[14]; and secondly, the private building strategies attested to by a vast aristocratic residence—namely, the Villa of Diomedes.[15] Instead a systematic approach was applied to an extended area of Pompeii on a large scale. That allowed to obtain a surveying and recording all the post-seismic repairs. We selected *Regio* VII, with its ten blocks, to the east of the forum (*insulae* n°1, 2, 3, 4, 9, 10, 11, 12, 13, and 14). It covers an area of 5 ha, around 15% of the discovered site, and constitutes a representative sample comprising public and religious buildings, houses, shops, and industries.[16] Such an approach to an "urban fragment" allows us to identify the various types of repairs and to compare the reconstruction conditions in the various functional categories of the city.

4.2.1 Distinguishing Between Post-Earthquake Operations: Reconstructions, Repairs, and Reinforcements

Most masonry structures of a monumental character belong to the specialized type of buildings that, by their nature, do not show a clear structural behaviour as a whole and, in this case, the assessment of seismic safety. The surviving structures' lesions, the original walls' out-of-plumpness, and the collapse profiles are the only valuable clues for reconstructing the most severe events that affected the archaeological area. These signs also make it possible to appreciate the quality of the constructions and the functional attention paid in the past to the solidity of the buildings. For structural analysis, there is, unfortunately, no shortage of opportunities for study, which arise dramatically after every disaster. Masonry structures are complex systems that require detailed knowledge and information regarding their response under seismic shaking. In structural analysis, one should consider these signs/traces/cracked squares that also show damaged buildings. In this context, damage pattern on ancient buildings can represent models for modern ones. Looking at an emblematic case such as Pompeii, we have several examples of repair and damage patterns and relevant information to prevent or repair. In our study in *Regio* VII, we counted 255 post-seismic interventions, including 184 partial repairs and 71 reconstructions.

It is repairing a building, or part of a building, involves replacing a damaged structure whose original configuration can be reproduced more or less faithfully in the post-seismic operation. A distinction should be made according to the scale of the repair and its relationship to the previous structural system. If the repair concerns the entire system, it is a wholesale reconstruction.

The other situation is a partial repair, i.e. only the portion of an element. In this case, three main operations can be identified, according to which elements are damaged, ranked in descending order according to the number of cases recognised in the survey of *Regio* VII:

- The restoration of jambs or architraves of openings (58 cases)
- The restoration of wall corners (34 cases)
- The filling of cracks in elevations, arches, or vaults (42 cases).

All these operations aim to restore the structural completeness of the building by intervening in areas considered most vulnerable and prone to

[14] Filocamo et al. (2018); Dessales et al. (2019).

[15] Dessales (2020a).

[16] Dessales (2022).

damage due to stress build-up. The technique of connecting surviving portions by partial repair is assimilated to "*scuci e cuci*" interventions, using the sewing metaphor (*unstitching and stitching*).[17]

Distinct from repairs, reinforcements are another type of operation, which are not carried out on the damaged elements but involve adding another element. The main function of these punctual strengthening devices is to counteract the horizontal force and the rotation of the façade thanks to their sheer mass. These solutions are also introduced to counteract the lateral forces exerted by roofs, arches, and vaults.[18] The reinforcements can be distinguished in:

- Buttresses (6 cases)
- Arch buttresses (4 cases)
- Filling of openings (2 cases)
- Wall redoublings (2 cases).

4.2.2 Identifying a Repair in a Damaged Building

In Pompeii, only ancient and not contemporary repairs were taken into account, as seismic episodes can be linked to events like the one in Irpinia (1980).[19] To identify them, several methods have been developed. The first one is the stratigraphic analysis applied to elevations, which follows well-known and widespread methods in the study of historical buildings and archaeoseismology.[20] Identifying the physical relationships between the building elements (covering: covers/is covered by; support: supports/leans against; cut: cuts/is cut by; filler: fills/is filled by/links: is linked to) and their chronological relationships allows to investigate the material structure of a building and to highlight all the phases of its life. The distinctive criteria used to determine the different stratigraphic units of a built element are based on four main characteristics:

- Nature of the building materials
- Shapes and finishing techniques of the materials
- Dimensions of the constituent elements
- Continuity of the surface (facing or plaster).

The second method used is characterizing the construction techniques and determining their chrono-typology, within the context of a single monument and in relation to the other buildings on the site. Various characteristics have been taken into account to identify the elevation techniques: general data on the wall structure (morphology in plan and section, static function, section) and more detailed data on the constituent parts (facing, core, or homogeneous structure). This approach considers two essential aspects: the nature of the building materials and the methods of putting them in place. The analysis is completed by a detailed description of the facing, which looks specifically at the layout of the materials:

- Shape, finishing, and hardness of the materials (stone or terracotta)
- Indicative measurements of the materials
- Morphology of the joints
- Presence of tool marks
- Presence of bonds between elements (clamp, dowel)
- Morphology of the courses (layout of the elements in facing and section, presence of levelling courses or wedges).

Moreover, particular attention is accorded to the structure of the vertical and horizontal reinforcements about the walls facing: quoins, piers, opening frames, belt courses, and eventual crowning.

In the case of Pompeii, the type of materials and the quality of the mortar allow us to distinguish ancient repairs from modern restorations, particularly those carried out from the end of the eighteenth century, during the first excavations of

[17] Doglioni (2018, 30).

[18] Ferrigni et al. (2005); Giuliani (2011); Arrighetti (2015, 101–103).

[19] Zevi (1982); Adam (1989).

[20] Arrighetti (2015, 95–100); Arrighetti (2018).

Fig. 4.1 Example of a modern repair in Pompeii, Villa of Diomedes: an arch buttress built in the 1770s (Photo T. Crognier, 2015)

the city.[21] In fact, buttresses, doubling of walls, or filling of openings, carried out to reduce the vulnerability of the elevations, were frequent at that time (Fig. 4.1). The presence of lapilli, recovered while excavating the buildings, is a good indicator of a mortar made in the Bourbon period, mainly at the end of the eighteenth century and the first half of the nineteenth century. The stratigraphic study and the characterization of the materials, in particular bricks and mortar, are the only methods for distinguishing a repair carried out in the Roman period.

In order to carry out the inventory of the ancient repairs in the field, two complementary tools were developed: firstly, a database designed to record the different types of repair and reinforcement and secondly, a mapping interface to

[21] OPUS database ("*Outil Pour Unités Stratigraphiques construites*"; Tools for Constructed Stratigraphic Units) developed to record the stratigraphic units and employed during the campaigns conducted on the Villa of Diomedes in Pompeii between 2012 and 2016. See the case study of the Villa of Diomedes: Dessales and Monier 2020, 82–90.

map these interventions in the ancient city of Pompeii.[22]

4.3 A Specific Database for the Analysis of the Repairs: OPUR

4.3.1 Technical Aspects and Interface

The OPUR database ("*Outil Pour Unités de Réparation*"; Tools for Repair Units) was designed for the analysis of the repairs identified in Pompeii,[23] following the structural organization of the OPUS database ("*Outil Pour Unités Stratigraphiques construites*"; Tools for Constructed Stratigraphic Units), developed earlier in the framework of various archaeological projects on Roman monuments, in particular the Villa of Diomedes, in Pompeii.[24] The OPUR database was developed at the same time as the observations were carried out in the *Regio* VII of Pompeii in 2016 and 2017 (6 weeks in total, with a first test on the Building of Eumachia). OPUR users worked on-site using PC Tablets to record the 255 repairs in the ten selected *insulae* of *Regio* VII. The data are archived, checked, and merged, updating a shared version. At the end of the field missions, 1500 photographs produced during the field campaigns were added to the database, and the full version was available through different levels of accessibility.

The OPUR database was designed using the software FileMaker and deposited on the remote server available via the French Research Infrastructure Huma-Num, a project dedicated to digital technologies in the field of the humanities and social sciences. The database is available for free from the website RECAP, in three languages (French, English, Italian).[25] The database can offer several additional functionalities, such as image management. A catalogue in PDF format can be automatically produced, selecting the data and images to be included.[26]

4.3.2 OPUR Database Content

OPUR database describes each repair according to a system of indexing and document analysis developed in collaboration with the structural engineers associated with the RECAP project. A predefined list of values is provided to ready-to-use on-site, which can be consulted either through the RECAP database or by consulting the paper record sheet (Fig. 4.2), particularly suitable for the analysis of elevations (as walls, buttresses, columns, and pillars) and openings (doors, windows, and niches). The paper record sheet allows us to test and adapt the method in the field, whether or not the information is intended to be recorded in the database.

OPUR consists of four distinct parts (Fig. 4.3):

- Identifying and locating the repair
- Identifying the damage
- Characterizing the nature of the repair
- Indicating chronological relations.

4.3.3 Repair and Damage

An identifier number and location are assigned to a repair. In the specific case of Pompeii, a distinction was made between *Regio*, *Insula*, and property number (*civico*). However, in the version suitable to any archaeological site, available for download, a georeferenced reference (x, y, z) is linked to each repair, allowing a study of the spatial distribution.

Each repair is defined by its technique and position within a construction element. A stratigraphic method was applied, such that two units present in the same elevation, but without any point of contact, were considered as two different repairs (indicated as contemporary in the chronological relationships).

[22] For a more in-depth methodological presentation of these tools, see Dessales et al. (2022).

[23] Dessales and Tricoche (2018).

[24] Tricoche (2020).

[25] Recap (2005).

[26] Catalogue published in Dessales (2022, 258–285).

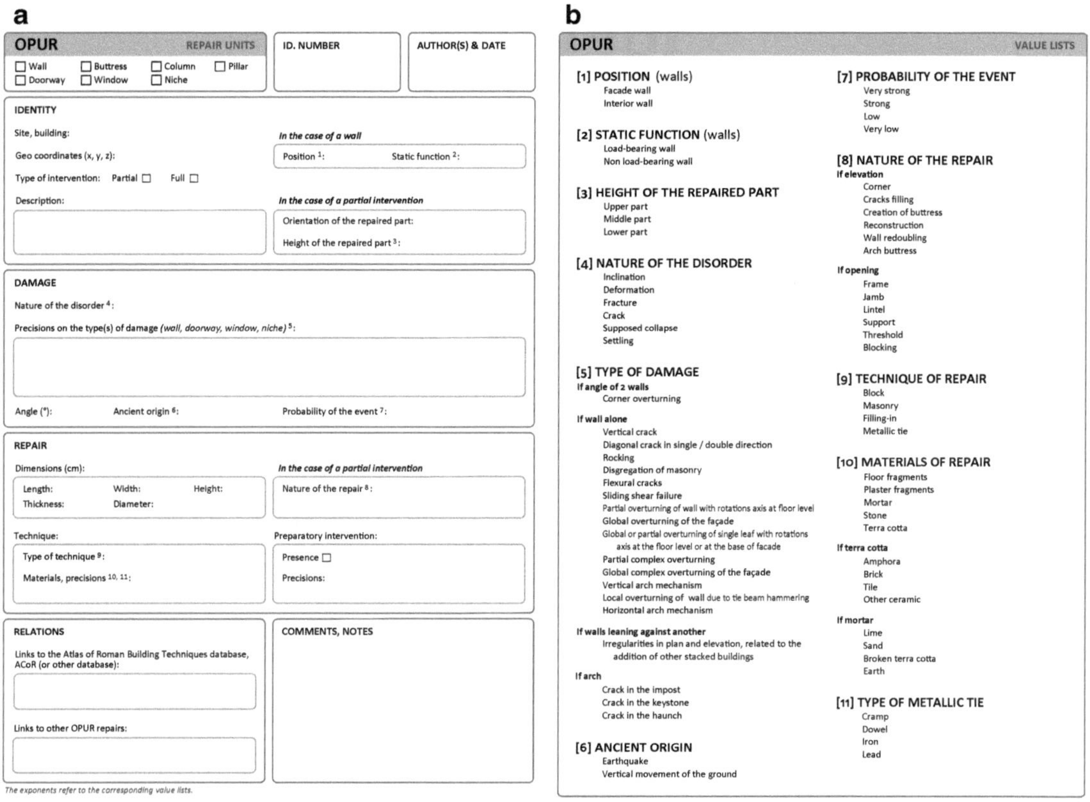

a

OPUR REPAIR UNITS
☐ Wall ☐ Buttress ☐ Column ☐ Pillar
☐ Doorway ☐ Window ☐ Niche

ID. NUMBER

AUTHOR(S) & DATE

IDENTITY
Site, building:
Geo coordinates (x, y, z):
Type of intervention: Partial ☐ Full ☐
Description:
In the case of a wall
Position [1]: Static function [2]:
In the case of a partial intervention
Orientation of the repaired part:
Height of the repaired part [3]:

DAMAGE
Nature of the disorder [4]:
Precisions on the type(s) of damage *(wall, doorway, window, niche)* [5]:
Angle (°): Ancient origin [6]: Probability of the event [7]:

REPAIR
Dimensions (cm):
Length: Width: Height:
Thickness: Diameter:
In the case of a partial intervention
Nature of the repair [8]:
Technique:
Type of technique [9]:
Materials, precisions [10, 11]:
Preparatory intervention:
Presence ☐
Precisions:

RELATIONS
Links to the Atlas of Roman Building Techniques database, ACoR (or other database):
Links to other OPUR repairs:

COMMENTS, NOTES

The exponents refer to the corresponding value lists.

b

OPUR VALUE LISTS

[1] POSITION (walls)
Facade wall
Interior wall

[2] STATIC FUNCTION (walls)
Load-bearing wall
Non load-bearing wall

[3] HEIGHT OF THE REPAIRED PART
Upper part
Middle part
Lower part

[4] NATURE OF THE DISORDER
Inclination
Deformation
Fracture
Crack
Supposed collapse
Settling

[5] TYPE OF DAMAGE
If angle of 2 walls
Corner overturning
If wall alone
Vertical crack
Diagonal crack in single / double direction
Rocking
Disgregation of masonry
Flexural cracks
Sliding shear failure
Partial overturning of wall with rotations axis at floor level
Global overturning of the façade
Global or partial overturning of single leaf with rotations axis at the floor level or at the base of facade
Partial complex overturning
Global complex overturning of the façade
Vertical arch mechanism
Local overturning of wall due to tie beam hammering
Horizontal arch mechanism
If walls leaning against another
Irregularities in plan and elevation, related to the addition of other stacked buildings
If arch
Crack in the impost
Crack in the keystone
Crack in the haunch

[6] ANCIENT ORIGIN
Earthquake
Vertical movement of the ground

[7] PROBABILITY OF THE EVENT
Very strong
Strong
Low
Very low

[8] NATURE OF THE REPAIR
If elevation
Corner
Cracks filling
Creation of buttress
Reconstruction
Wall redoubling
Arch buttress
If opening
Frame
Jamb
Lintel
Support
Threshold
Blocking

[9] TECHNIQUE OF REPAIR
Block
Masonry
Filling-in
Metallic tie

[10] MATERIALS OF REPAIR
Floor fragments
Plaster fragments
Mortar
Stone
Terra cotta
If terra cotta
Amphora
Brick
Tile
Other ceramic
If mortar
Lime
Sand
Broken terra cotta
Earth

[11] TYPE OF METALLIC TIE
Cramp
Dowel
Iron
Lead

Fig. 4.2 Paper record sheet of the OPUR database (A. Tricoche)

OPUR focuses on the repairs as indirect effects of earthquakes, allowing to go back to direct effects, based on the characterization of the related damage.[27] A multidisciplinary team was involved in determining the nature of the disorder: inclination, deformation, fracture, crack, supposed collapse, and settling. From this observation, the type of damage is deduced, and its potential link with an earthquake is assessed. In *Regio* VII, five local damage mechanisms have been identified: corner overturning, overturning of the façade, vertical arch mechanism, and horizontal arch mechanism. They are the result of a poor connection between the façade and the other portions of the buildings (walls, floors).

In contrast, in cases where the connections between the walls and floors were better, the damage is concentrated in the plane of the wall. It is due to three global mechanisms: vertical crack, diagonal crack in a single direction, and diagonal crack in a double direction. Some damage can also be observed at the impost (springer) or the haunch of arches or vaults as a response to the increased vertical load caused by the seismic shake.

4.3.4 Building Techniques and Chronological Relations

A repair is characterized by the technique used (blocks, masonry, metallic ties). For each of these

[27] The types of damage are listed with reference to the codification proposed in the *Linee guide per la valutazione e riduzione del rischio sismico del patrimonio culturale*: Ministero per i Beni e le Attività Culturali (2010); Papa and Di Pasquale (2011) moreover, a schematic illustration was provided for each type by Giuseppina De Martino and Francesca Autiero, under the direction of Andrea Prota (Università degli Studi di Napoli Federico II, Dipartimento di Strutture per l''Ingegneria e l''Architettura).

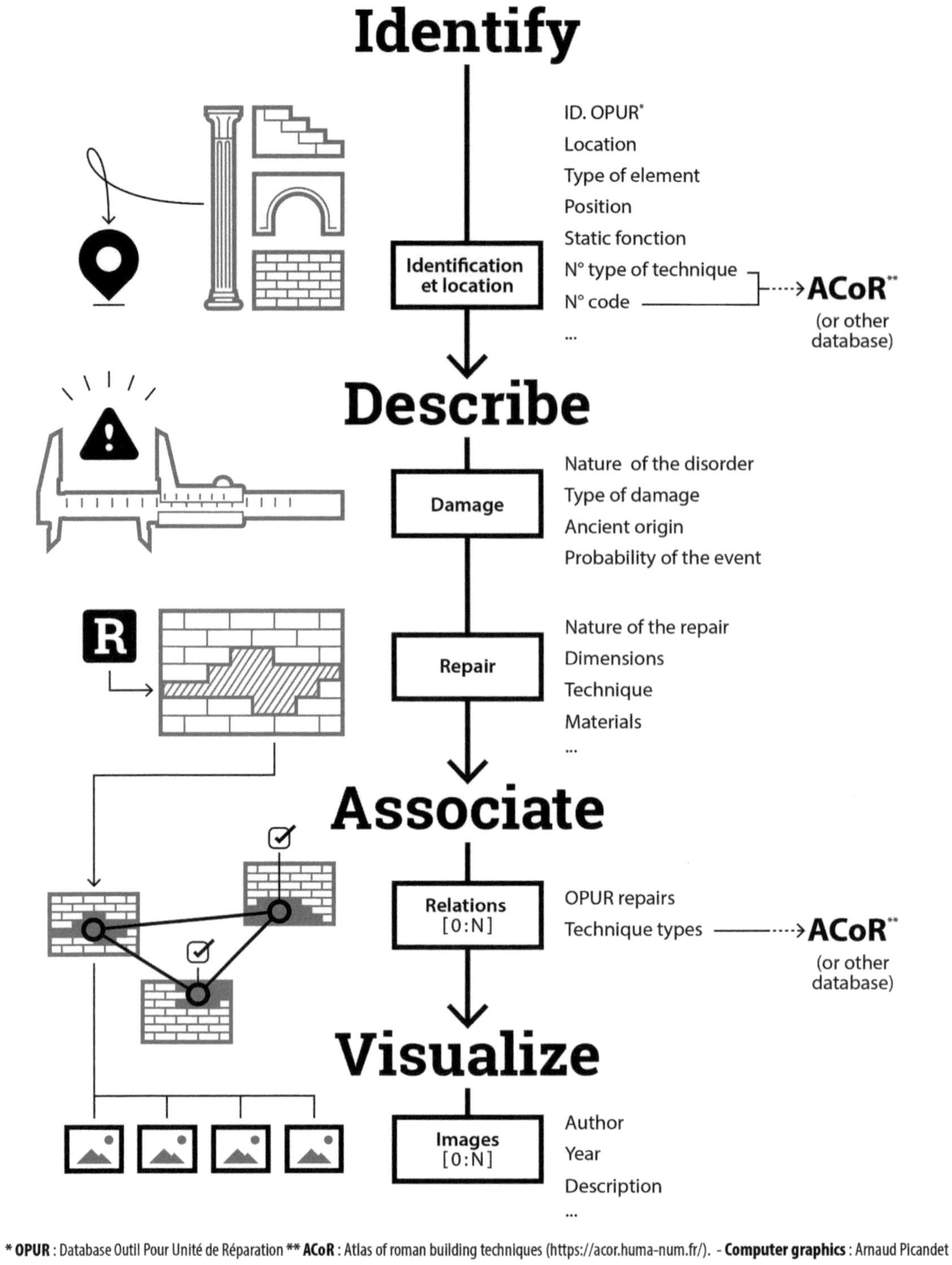

Fig. 4.3 Diagram of the organization of the OPUR database (A. Picandet, A. Tricoche)

three categories, OPUR proposes a list of identified materials used for the repair, which can be enriched according to the specificities of each archaeological site. In addition, any other external reference can be indicated to link the repair with a parallel indexing system on building techniques, using an identifier number.[28] Thus, for the study of *Regio* VII in Pompeii, we have referred to the database ACoR, developed to record Roman building techniques.[29] This international project led to the creation of a controlled, hierarchized vocabulary for defining and characterizing building techniques in the Roman world in the form of a new trilingual thesaurus (French, English, and Italian).[30] Moreover, it allows consulting the different types of techniques identified. In the case of elevations, the criteria used to define the types are as follows:

- Section of elevations (homogeneous or with facings)
- Size of facing materials (small or large units)
- Shape of facing materials or homogeneous structure
- Layout of facing materials or homogeneous structure (without courses, or horizontal, inclined, undulating, concave, 45° courses, courses in chequer or/and herringbone pattern, roughly backfilled concrete, concrete backfilled in successive layers).

The types established according to these different criteria have a five-digit numbering, determined automatically, and reported in the OPUR database if the repair corresponds to a defined technical type listed in ACoR. Similarly, the kind of technique that characterizes the initially damaged structure is indicated (Fig. 4.4).

Based on these relationships, it is possible to develop a statistical approach to identify the most damaged types of techniques and those commonly used in the case of repairs. For example, regarding the repair of walls, as one of the most damaged elements, just over half of all repairs recorded (135 cases of partial repairs or total reconstructions), facings and quoins characterize 27% of the repairs in ceramic building materials (ACoR type No. 12-05-002), and 21% by facings in irregular rubble masonry and quoins that alternate small rectangular blocks and terracotta (ACoR type No. 12-04-008) (Fig. 4.4). Such a result coincides with the conclusions regarding the diffusion of terracotta in the reconstruction of Pompeii.[31] On the other hand, the technique that appears to be the most damaged (43.8%), compared to the total of 226 cases for which it was possible to identify the previous damaged structure, is a masonry of irregular rubble, with large travertine blocks in quoins (ACoR type No. 12-04-017 technique).

Moreover, an internal link is proposed within the framework of the OPUR database between the various repair units. Indeed, a repair may be posterior, anterior, or contemporary to one or more other repairs. In the case of Pompeii, the identification of 11 repairs successive to previous repairs, all related to a seismic episode, tends to prove the succession of several earthquakes in the final years of the city.[32]

4.4 The Cartography and Quantification of Repairs

A systematic mapping of the various repair units recorded in the OPUR database is associated with a Geographic Information System (GIS) (ArcGIS v.10). It allows us to undertake spatial and quantitative analyses of post-seismic interventions and seismic impacts at the scale that we have considered. From a technical point of view, we aimed to create an effective registration interface to be implemented on-site, useful also for the methodology to define the elements that needed to be surveyed. In fact, it was necessary for *Regio* VII to determine the various criteria for mapping the interventions in addition to the information recorded in the OPUR database.

28 See, for example, Arrighetti (2015, 99–106).

29 Acor (2016).

30 Camporeale et al. forthcoming.

31 Dessales (2016).

32 Dessales (2020a, 427–446).

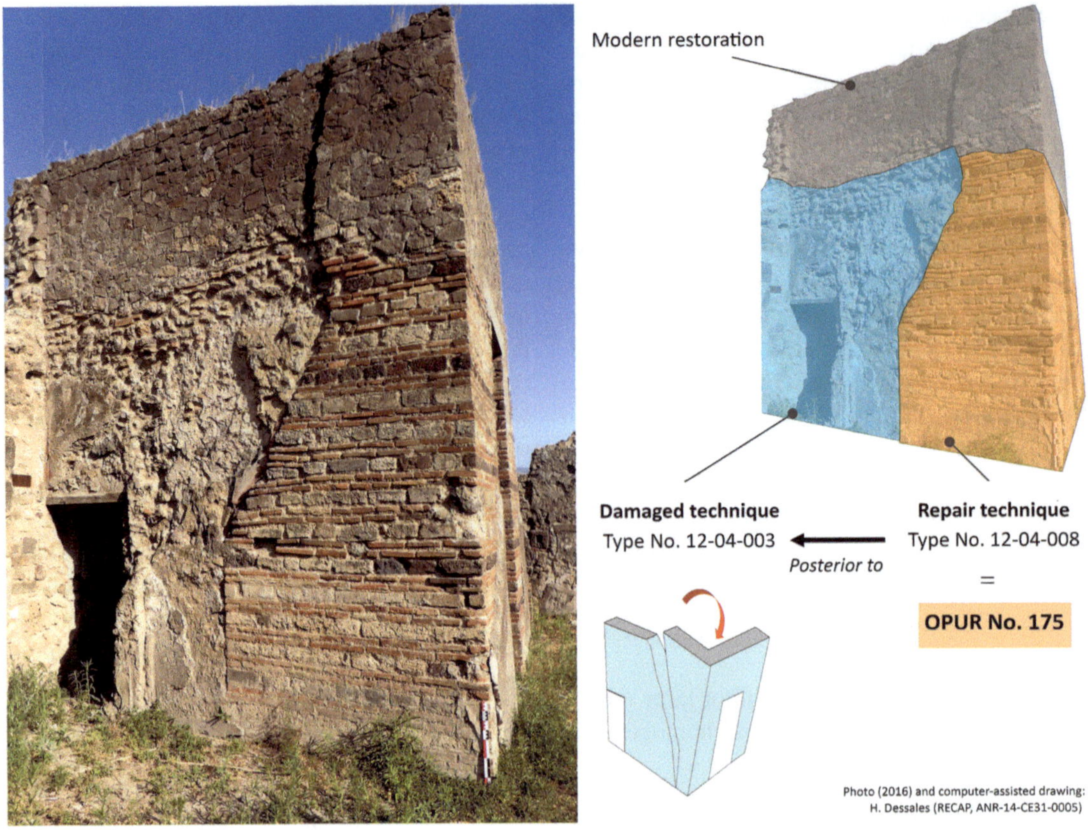

Fig. 4.4 Example of an OPUR unit identification (*Casa di Ganimede*, VII 13, 4): repair consecutive to a corner overturning, OPUR unit no. 175 (H. Dessales)

4.4.1 Technical Aspects of the Field Data Acquisition

The project covered on-site both urban and architectural scales for the vast area (10 *insulae*–5 ha with more than 300 properties). For this reason, we applied Collector for ArcGIS,[33] an accessible geographic data collection application for mobile devices developed by ESRI (Fig. 4.5). The system is provided with a photographic sensor, so the photographs are attached directly to the entities surveyed. Instead, the data collected was exported in SQLite format and synchronized with the GIS project.[34] Unfortunately, due to their peculiarities, some indoor and urban-scale cases did not allow the use of a Global Navigation Satellite System. To solve these limitations, a 1/50 plan adapted to a multi-scale operation was produced,[35] coupled with an aerial photograph of the area and other information that has allowed us to associate the civic numbers with a geocoding system on site.[36] To obtain a dynamic base map, which allowed offering several levels of detail depending on the display scale, the procedure followed two steps. First, the georeferenced

[33] Collector for ArcGIS is based on the editing pre-configured geographical entities (geometry and attribute table) on a background map loaded into the mobile device

[34] They can be post-synchronized also in an offline mode.

[35] Thanks to the collaboration with the Archaeological Park of Pompeii, this plan was based on the documentation produced in the framework of the "Nuova cartografia informatizzata georiferita" initiated as part of the "Grande Progetto Pompei (Piano della Conoscenza)".

[36] Morichi (2018).

Fig. 4.5 Field recording interface based on a 1/50 plan (collector for ArcGIS) (J. Cavero)

vector plans were imported into the GIS and combined into several feature classes. Second, the feature classes were classified by floor after conducting a topological simplification to remove overlapping areas. Then published as a tile map service, the base map was available on a PC tablet. In this way, the mapping of the repairs was made in the field, at the same time as the data recording in OPUR, during a single 3 week mission in 2017.

4.4.2 Recording and Mapping Repairs

As a first relief, it was necessary to define the elements relevant to post-seismic repairs and reconstructions to be surveyed, essential for the data inserted in GIS, and to ensure its complementarity with the OPUR database. The aim was to be able to study, beyond the individual study of each repair, their impact on the urban scale. Four relevant categories of elements to post-seismic repairs and constructions were identified, each element being part of a geographical, and therefore geometric, expression in the base map (Fig. 4.6):

- Repairs
- Reconstructions associated with a repair
- Constructions associated with a repair
- Post-seismic building processes.

Repairs are the individual building intervention units following seismic damage, whose seismic origin is probable. All repairs correspond to a unique record in the database, unlike the other three categories, which have only been mapped. For each element, the ID generated by the database was reported to connect the two recording systems. Repairs with sufficient surface area to be mapped at 1/50 are recorded in a polygonal shape (characterized by their length, orientation, and

Fig. 4.6 Recorded post-earthquake interventions in *Regio* VII (J. Cavero)

volume). On the other hand, crack fillings, block repairs, and interventions on columns or pillars. Reconstructions associated with a repair are those elements that belong to the same building process, a specific repair unit, involving the identical reconstruction of previous elements but excluding transformations or redevelopments of structures (e.g. modification of a house plan). Constructions associated with a repair are defined similarly, but in this case, it is not possible to specify whether they are an identical reconstruction of previous elements. Post-seismic building processes are those elements whose construction techniques and materials are characteristic of the last of the city.[37] They feature new constructions and exclude any integration of previous structures. The last three categories are recorded in a polygonal shape. This distinction is useful to develop a quantitative and spatial approach of post-seismic (re)constructions, which is not limited to repair units and allows for reasoning in terms of building processes.

4.4.3 Quantitative Analysis at the Urban Scale

A spatial database enables many observations and cross-referencing analysis of information. For example, the distribution of different construction techniques used for repairs, mentioned in the previous section on the database, can be spatially translated. Data relative to the estimated volumes of materials employed in repairs or (re)-constructions can also be processed.

The analysis of the post-seismic repair interventions recorded is carried out using a multi-scalar approach to define their organization clearly and propose a mapping of the impacts of seismic activity on the site, following three different scales:

- The building unit, to define the vulnerability of each building and the relevance of the repair

[37] Maiuri (1942); Dessales (2016, 2022); Ruggieri (2019). This is the case, for example, with some of the public buildings in the forum (VII 8), which have been rebuilt with terracotta masonry of cut tiles.

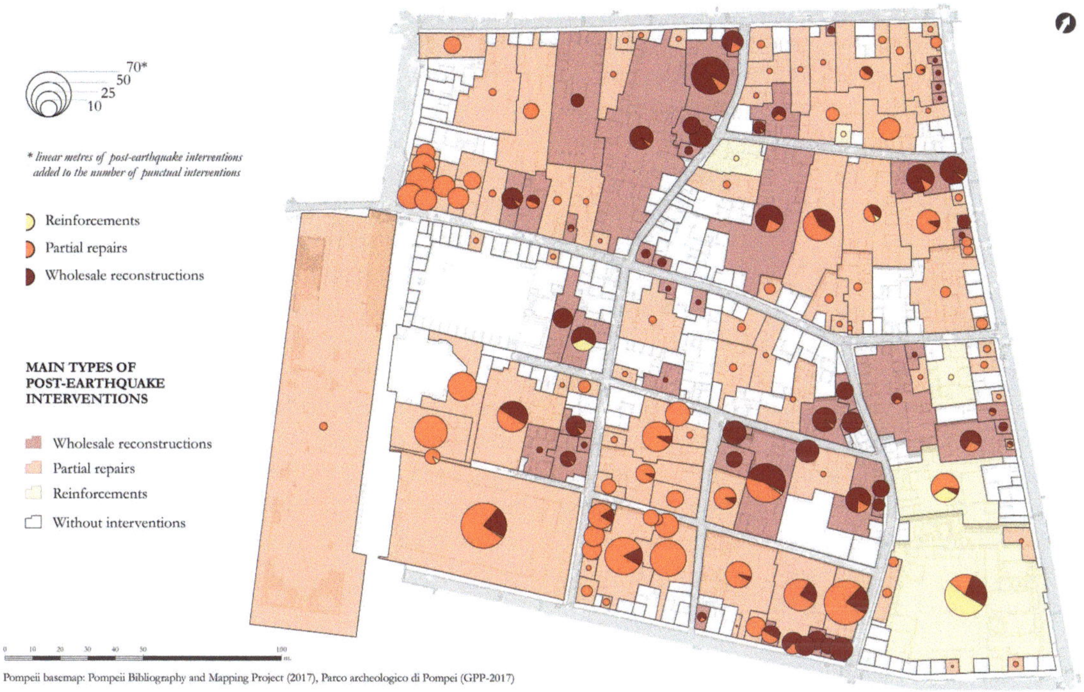

Fig. 4.7 Typology by parcel of the post-earthquake interventions surveyed in *Regio* VII (J. Cavero)

- The parcel,[38] in order to define the connection between the different constructions[39]
- The entire city, to compare the function of buildings and the distribution of the various construction sites following the earthquake.

If we consider, for example, the scale of the parcel, it allows us to highlight the types of space most impacted by the earthquake. In Regio VII, 310 parcels were identified, which can be grouped into five functional categories; they are extremely variable in size, with roughly an average of 1000 m^2 for public buildings (not counting the Forum, which measures 5000 m^2), while smaller shops may be as small as 9 m^2 (Fig. 4.7). One hundred fifty-eight parcels, i.e., more than half of all plots, have undergone at least one post-seismic intervention. Most were mainly affected by partial repairs (98 out of 158). Wholesale reconstructions are the majority for 55 of these cases, distributed in almost all the insulae. Reinforcements stand out only in 5 parcels, concentrated essentially in insula VII 1, around the Stabian Baths.

Such an approach allows us to reflect on the spatiality of seismic risk in Pompeii. The study conducted on the scale of the city enables us to quantify the reconstruction sites and, in this way, to quantify the damage caused by the seismic impacts, and to determine the macroseismic intensity, according to the different scales available (EMS 98, for example).[40] Preliminary studies indicated an intensity of IX in the MCS scale for the 62/63 CE earthquake,[41] with a magnitude evaluated at five.[42] Through our study based on a

[38] Defined as a built unit within a block (*insula*), circumscribed by walls, and therefore without communication with other units.

[39] In particular, those relative to topographical levels, that exist in the city due to the presence of underground spaces in the various buildings, concentrated mainly to the east of *Regiones* VI and VII (in particular, VII 2).

[40] Grünthal and Levret (2001).

[41] Adam (1989); Guidoboni et al. (2018).

[42] Marturano et al. (1998); Cubellis et al. (2007).

large scale and the quantification of the repaired and destroyed structures, we can propose a level of urban destruction of 20% with a degree IX MCS. On the EMS scale, the intensity would be IX, considering a damage level on grade D5 (vulnerability of the structures: level A/B). A preliminary assessment will need to be refined by further study.[43]

4.5 Conclusion

The analysis of post-earthquake repairs provides insight into the relationship between building techniques, economy, and risk. In the case of Pompeii, it highlights adaptation strategies adopted by Roman society in the wake of an earthquake by defining the typology of repair techniques used. As such, the variability of the techniques employed becomes apparent, depending on the nature of the damage and the category of the building in question. Through this approach, it is possible to evaluate the capacity of ancient societies to recover after a seismic event. In the absence of written evidence on the ways of rebuilding in the wake of an earthquake, this archaeology of the repairs is of primary importance to understand the rebuilding processes and the economic choices. Such an analysis, focused on the technique of the repairs, also provides some additional insight into the conception that ancient builders could have of reconstruction or maintenance. Indeed, beyond its material realization, a repair should include a knowledge questioning that not only aims to restore a state following a disorder but may also assess the origins of the damage and the ways to avoid it.[44]

Developed on Pompeii's unique laboratory, the protocol of identifying repairs could be extended to other ancient sites or historic buildings in prone-earthquake areas, regardless of the historical period considered. The OPUR protocol thus allows a multi-scalar approach, either at the urban scale of a large site or at the individual scale of a monument. Indeed, it has recently been successfully applied in the analysis of the mediaeval church of Sant'Agata del Mugello (Tuscany, Italy).[45] In this church, which served as a multidisciplinary laboratory for the study of seismicity in the Mugello basin, 80 repair units were identified and associated with damage mechanisms. Combining this data and historical records, thirteen successive building phases could be specified and dated. It appears that the church was deeply affected by two seismic events in 1542 and 1611. By highlighting the repairs and transformations that a building undergoes over time, such a study makes it possible to examine both the improvised restorations and the preventive techniques implemented to reduce architectural vulnerability. It can thus be the basis for a broader approach to built heritage.

Acknowledgments We express our warmest thanks to David A. Holmes for his accurate review of this paper. Our gratitude goes to the RECAP "*Regio* VII" team: Francesca Autiero, Guilhem Chapelin, Marina Covolan, Arnaud Coutelas, Thomas Crognier, Giuseppina De Martino, Marco Di Ludovico, Christophe Loiseau, Florence Monier, Arnaud Montabert, and Andrea Prota. Without the kind support of the team at the Archaeological Park of Pompeii, this programme might never have seen the light of day. We want to thank its director, Massimo Osanna, architecture officer Annamaria Mauro, archaeology officers Laura De Esposito, Bruno De Nigris, Giuseppe Scarpati.

References

Acor (2016) ACoR, atlas des techniques de la construction Romaine (atlas of Roman building techniques). https://acor.huma-num.fr/. Accessed on 15 Feb 2020

Adam JP (1989) Osservazioni tecniche sugli effetti del terremoto in Pompei del 62 d.C. In: Guidoboni E (ed) I terremoti prima del Mille in Italia e nell'area mediterranea. SGA, Bologna, pp 168–171

Alberti LB (1988) In: Rykwert J, Tavernor R, Leach N (eds) De re aedificatoria. On the art of building in ten books. The MIT Press, Cam-bridge Mass

Arrighetti A (2015) L'archeosismologia in architettura. Per un manuale. Florence University Press, Florence

Camporeale S, Dessales H, Tricoche A (n.d.) Forthcoming. In: Atlante delle tecniche della

43 Dessales (2022, 210–221).

44 On this culture of repairing, see Dessales (2020b).

45 Montabert et al. (2020).

costruzione romana. Manuale (Costruire nel mondo antico). Qasar, Rome
Cubellis E, Luongo G, Marturano A (2007) Seismic hazard assessment at Mt. Vesuvius: maximum expected magnitude. J Volcanol Geotherm Res 118:339–335
Davoine C, L'Héritier M, Péron d'Harcourt A (eds) (2019) Sarta tecta. De l'entretien à la conservation des édifices. Antiquité, Moyen Âge, début de la période moderne (BiAMA, 26). PUP, Aix-en-Provence
Dessales H (2012) Not built in a day: awareness of vulnerability and construction techniques in Roman times. In: Carvais R, Guillerme A, Nègre V, Sakarovitch J (eds) Nuts & bolts of construction history. Culture, technology, and society, vol 3. Picard, Paris, pp 471–477
Dessales H (2016) La produzione laterizia a Pompei. Adeguamento di un materiale e organizzazione dei cantieri urbani. In: Bukowiecki E, Volpe R, Wulf-Rheidt U (eds) Il laterizio nei cantieri imperiali. Roma e il Mediterraneo (Archeologia dell'architettura 20). Insegna del Giglio, Sesto Fiorentino, pp 81–89
Dessales H (ed) (2020a) The villa of Diomedes. The making of a Roman villa in Pompeii (histoire et archéologie; collection du Centre Jean Bérard 52). Hermann, Paris
Dessales H (2020b) Construction et risque sismique à l'époque romaine. Aedificare 7:45–76
Dessales H (ed) (2022) Ricostruire dopo un terremoto. Riparazioni antiche a Pompei (Collection du Centre Jean Bérard). Centre Jean Bérard, Naples
Dessales H, Monier F (2020) Excavations, removal, and restorations. In: Dessales H (ed) The villa of Diomedes. The making of a Roman villa in Pompeii (histoire et archéologie; collection du Centre Jean Bérard 52). Hermann, Paris, pp 77–90
Dessales H, Tricoche A (2018) Un database per studiare le riparazioni post-sismiche. Archeologia dell'architettura 23:19–24
Dessales H, Carlut J, Filocamo F (2019) L'entretien d'un aqueduc face au risque sismique: le cas de l'aqueduc du Serino (Italie). In Ronin M, Möller C (eds) Instandhaltung und Renovierung von Straßen und Wasserleitungen von der Zeit der römischen Republik bis zur Spätantike. Entretien et restauration des infrastructures routières et hydrauliques de l'époque républicaine à l'Antiquité tardive (Berliner Schriften zur Rechtsgeschichte 10): 135–160. Nomos, Berlin
Dessales H, Cavero J, Tricoche A (2022) Post-earthquake reconstruction: mapping and recording repairs in ancient Pompeii. In: D'Amico S, Venuti V (eds) Handbook of cultural heritage analysis: 1805–1821. Springer, Cham
Doglioni F (2018) Il dannegiamento sismico come processo. La lettura archeosismologica come strumento di prevenzione. Archeologia dell'architturа 22:25–38
Ferrigni F, Helly B, Mendes Victor H, Pierotti P, Rideaud A, Teves Costa P (2005) Ancient buildings and earthquakes. The local seismic culture approach: principles, methods, potentialities. Edipuglia, Bari
Filocamo F, Carlut J, Dessales H, Borensztajn S (2018) Roman builders facing the risk of disaster. Coupling archaeological and geochemical analyses of a section of the «Aqua Augusta» (the Roman Serino aqueduct, southern Italy). Archaeometry 60(3):915–932
Fröhlich T, Jacobelli L (eds) (1995) Archäologie und Seismologie. La zona vesuviana dal 62 al 79 D.-C. Problemi archeologici e sismologici. Biering und Brinkmann, Munich
Giuliani CF (2011) Provvedimenti antisismici nell'antichità. J Ancient Topogr 21:25–52
Grünthal G, Levret A (2001) L'échelle Macrosismique Européenne. European Macroseismic Scale 1998. Conseil de l'Europe, Luxembourg
Guidoboni E, Ebel JE (2009) Earthquakes and tsunamis in the past. A guide to techniques in historical seismology. Cambridge, Cambridge University Press
Guidoboni E, Ferrari G, Mariotti D, Comastri A, Tarabusi G, Sgattoni G, Valensise G (2018) CFTI4Med. In: Catalogue of strong earthquakes in Italy (461 BC–1997) and Mediterranean area (760 BC–1500). INGVA-SGA, Bologne. http://storing.ingv.it/cfti/cfti5/quake.php?50039IT>
Jusseret S, Sintubin M (eds) (2017) Minoan earthquakes. Breaking the myth through interdisciplinarity (Studies in archaeological sciences 5). Leuven University Press, Leuven
Jusseret S, Langohr C, Sintubin M (2013) Tracking earthquake archaeological evidence in late Minoan IIIB (~1300-1200 B.C.) Crete (Greece): a proof of concept. Bull Seismol Soc Am 103:3016–3043
L'Héritier M, Davoine C, Dessales H, Bernardi P (2018) Regular building maintenance and long-term conservation in ancient times. In: Wouters I, van de Voorde S, Bertels I (eds) Building knowledge, constructing histories: proceedings of the 6th international congress on construction history (6ICCH), Brussels, Belgium, July 9–13, 2018. CRC, Brussels, pp 31–38
Maiuri A (1942) L'ultima fase edilizia di Pompei. Istituto di studi romani, Spoleto
Marturano A, Rinaldis V (1998) Seismicity before the 79 AD Vesuvius eruption. In: Albore Livadie C, Ortolani F (eds) Il sistema uomo-ambiente tra passato e presente. Centro Universitario Europeo per I Beni Culturali, Ravello, pp 237–246
Ministero per i Beni e le Attività Culturali (2010) Linee guida per la valutazione e riduzione del rischio sismico del patrimonio culturale. Gangemi, Rome
Montabert A, Dessales H, Arrighetti A, Clément J, Lancieri M, Lyon-Caen H (2020) Tracing the seismic history of Sant'Agata del Mugello (Italy, Tuscany) through a cross-disciplinary approach. J Archaeol Sci Rep 33:102440
Monteix N (2011) Pompéi, observatoire privilégié d'une résilience urbaine inachevée? In: Djament G, Reghezza M (eds) Résiliences urbaines. Les villes face aux catastrophes. Éditions du manuscrit, Paris, pp 47–71
Morichi R (2018) Il disegno di una città antica: le scelte grafiche per una gestione dinamica delle

informazioni. In: Morichi R, Paone R, Sampaolo F (eds) Pompei. Nuova cartografia informatizzata georiferita. Arbor Sapientiae, Rome, pp 139–158
Papa S, Di Pasquale G (2011) Manuale per la compilazione della scheda per il rilievo del danno ai beni culturali, Chiese-Modello A-DC. Presidenza del Consiglio dei Ministri, Dipartimenti della Protezione Civile, Rome
Recap (2005) RECAP, Reconstruire après un séisme. Expériences antiques et innovations à Pompéi, http://recap.huma-num.fr/webpublic/. Accessed 15 Feb 2020
Rodriguez Pascua MA, Pérez-López R, Giner-Robles JL, Silva PG, Garduño-Monroy VH, Reicherter K (2011) A comprehensive classification of earthquake archaeological effects (EAE) in archaeoseismology: application to ancient remains of Roman and Mesoamerican cultures. Quat Int 242:20–30
Ruggieri N (2019) Pompei. Tecniche costruttive, vulnerabilità sismica, riparazioni e rinforzi al tempo d'll'eruzione del 79 d.C. Aracne, Roma
Savino S (2009) Nerone, Pompei e il terremoto del 63 d.C. In: Storchi Marino A, Mérola D (eds) Interventi imperiali in campo economico e sociale. Edipuglia, Bari, pp 225–244
Sintubin M (2017) Archaeoseismology. In: Jusseret S, Sintubin M (eds) Minoan earthquakes. Breaking the myth through interdisciplinarity (Studies in archaeological sciences 5). Leuven University Press, Leuven, pp 81–94
Tricoche A (2020) Database management system. In: Dessales H (ed) The Villa of Diomedes. The making of a Roman villa in Pompeii (Histoire et archéologie; Collection du Centre Jean Bérard 52). Hermann, Paris, pp 121–132
Varone A (2005) Convivere con i terremoti. La travagliata ricostruzione di Pompei dopo il terremoto del 62 d.C. alla luce delle nuove scoperte. In: Mols STAM, Moormann EM (eds) Omni pede stare. Saggi architettonici e circumvesuviani in memoriam Jos de Waele. "L'Erma" di Bretschneider, Naples, pp 315–324
Zevi F (1982) Pompei. In: Sisma 1980. Effetti sul patrimonio artistico della Campania e della Basilicata. Campania (Bolletino d'Arte Supplement 2). Istituto Poligrafico e zecca dello stato, Rome, pp 75–82

5 Application of Geological Techniques to Identify Earthquakes Affecting Archaeological Sites: The Example of the 2011 Lorca Earthquake (SE Spain)

M. A. Rodríguez-Pascua, R. Pérez-López, P. G. Silva, J. L. Giner-Robles, M. A. Perucha, J. Élez, T. Bardají, and E. Roquero

Abstract

The main uncertainty when studying the destruction horizon of an archaeological site or a heritage building is to know which trigger mechanism generated these deformations. This is the case of archaeoseismology, which needs to have a method to be able to discriminate whether the ruin of a given site was caused by simple abandonment or by an earthquake. The structural geology discipline studies, among other aspects, the deformations in rocks in relation to the intrinsic properties (fabric and mineralogy) and the tectonic parameters. This technique has demonstrated its utility to identify deformations of seismic origin by measuring the orientation and clustering the deformation to find spatial patterns of ground shaking and building response. The passage of the seismic waves from an earthquake depends on the seismic source and the epicentral location, and the so-called seismic ray arrives to the archaeological site with a trending which is measurable. The 2011 Lorca earthquake (Mw 5.2, SE of Spain) was a good example of the application of this novel methodology to determine the pattern of deformation affecting the buildings. During this earthquake, a lot of relevant and ancient buildings were affected by the seismic ray, giving a good opportunity for calibrating the spatial patterns of building deformation with the seismological parameters and geological features of the seismic source. Several Earthquake Archaeological Effects (EAEs) were mapped and compared with the orientation obtained from a seismic accelerometer of the Spanish National Seismic Monitoring Network, located in the city hall (in the historic downtown of Lorca). The ground-shaking movement, obtained from the application of this novel technique is NW-SE trending, and it agrees with the main orientation obtained from the seismic accelerometer.

M. A. Rodríguez-Pascua (✉) · R. Pérez-López · M. A. Perucha
Instituto Geológico y Minero de España (IGME), Madrid, Spain
e-mail: ma.rodriguez@igme.es

P. G. Silva · J. Élez
Department of Geología, Escuela Politécnica Superior de Ávila, Universidad Salamanca, Hornos Caleros, Ávila, Spain

J. L. Giner-Robles
Department of Geología, Facultad de Ciencias, Universidad Autónoma de Madrid, Cantoblanco. Tres Cantos, Madrid, Spain

T. Bardají
U.D. Geología, Universidad de Alcalá, Alcalá de Henares, Madrid, Spain

E. Roquero
Department of de Edafología, Universidad Politécnica de Madrid, Madrid, Spain

L. Pecchioli (ed.), *Archaeoseismology*, Natural Science in Archaeology,
https://doi.org/10.1007/978-3-031-28303-1_5

Keywords

Earthquake archaeological effects (EAEs) · 2011 Lorca earthquake · Cultural heritage · Structural geological analysis · Oriented seismic deformations

5.1 Introduction

Quantification is necessary to differentiate the effects of an earthquake from other non-seismic effects on archaeological sites and heritage buildings. This is important both for recognizing historically unrecorded earthquakes and destructive historical earthquakes, for which the seismic source is still under debate. The branch of geology that addresses the study of deformations in rocks is Structural Geology. In this way, different authors (Rodríguez-Pascua et al. 2011, 2012; Giner-Robles et al. 2009, 2011, 2012, 2018, 2022) proposed a new methodology based on the strain techniques to estimate patterns of seismic deformation affecting buildings and man-made constructions. The hypothesis maintains that the seismic waves arrive oriented to a village, and the anisotropic ground shaking by the passing of the surface seismic waves generates a measurable deformation pattern of damage affecting buildings. This methodology was applied to the Roman city of Baelo Claudia (S of Spain) (Silva et al. 2005, 2009, 2016), and two potential new earthquakes between the first and second centuries CE were recognized from the comprehensive study and cartography of different building damages, according to the classification proposed by Rodríguez-Pascua et al. (2011). In 2011, a destructive earthquake took place in Lorca (Mw 5.2, Murcia, SE Spain), a historical village with a rich presence of buildings built during different cultural periods since the tenth century CE. All of them were affected by the ground shaking, from small effects on the facade to damage close to the ruin and affecting the building foundation. These deformations were classified and mapped, obtaining a seismic spatial pattern for the building damage (Rodríguez-Pascua et al. 2012; Giner-Robles et al. 2012). Data collected from an accelerometer installed in the basement of the Lorca Town Hall, allows comparison between the anisotropic seismic vibration with this spatial pattern. Hence, instrumental data could validate the use of this technique based on the mapping of Earthquakes Archaeological Effects (EAEs, Rodríguez-Pascua et al. 2011), to be applied in other archaeological places where destructive earthquakes have occurred.

This paper describes the EAEs observed in the city of Lorca after the 2011 earthquake and the results of the geological structural analysis applied on it. The methodology is divided into three stages (Rodríguez-Pascua et al. 2011, 2012; Giner-Robles et al. 2012, 2018, 2022): (1) identification, location, and classification of the building damage; (2) quantification of the deformation through geological structural analysis, and (3) mapping the spatial damage-pattern, assignment of the seismic origin, and the incorporation of the EAEs into seismic databases and catalogues (Fig. 5.1).

5.2 The 2011 Lorca Earthquake and the EAEs

The seismic series of Lorca began on 11 May, at 15:05 UTC (local time 17:05), with an earthquake of Mw = 4.6 (2 km depth). Later, at 16:47 UTC took place the main earthquake of Mw = 5.2 (3 km depth), and 0.41 g of ground peak acceleration (http://www.ign.es), that produced serious damages in Lorca (Fig. 5.2). The second earthquake killed nine people, 324 people were injured, and about 20,000 were displaced from their homes. The epicentre was located at the NE of the city of Lorca, a few kilometres to the city on the Alhama de Murcia Fault (FAM) (Martínez-Díaz et al. 2011, 2012; Ortuño et al. 2012; Pérez-López et al. 2012a, b; Martínez-Solares & Mezcua, 2002) (Fig. 5.1). The seismic source was the segment of the FAM located between Lorca and Totana cities (Martínez-Díaz et al. 2012). This earthquake generated much damage affecting all of the Cultural Heritage buildings and even nowadays, damage can be recognized on

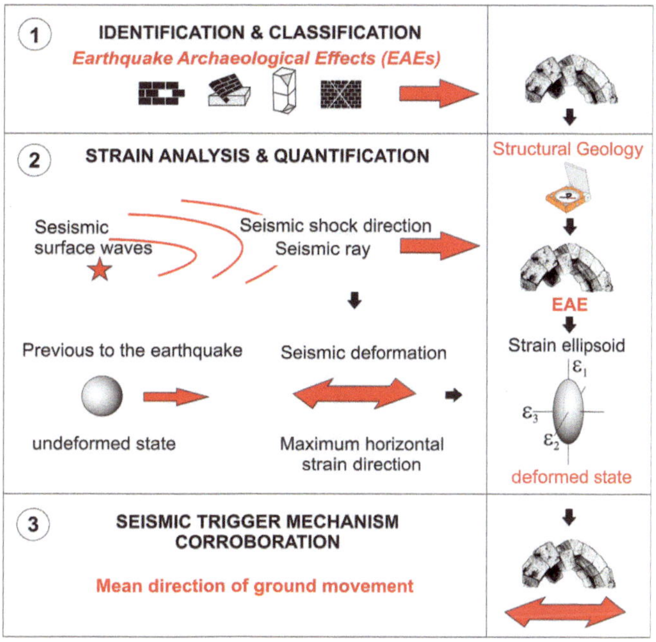

Fig. 5.1 The methodological flowchart and different steps of applying geological structural analysis to seismic deformations generated in archaeological sites

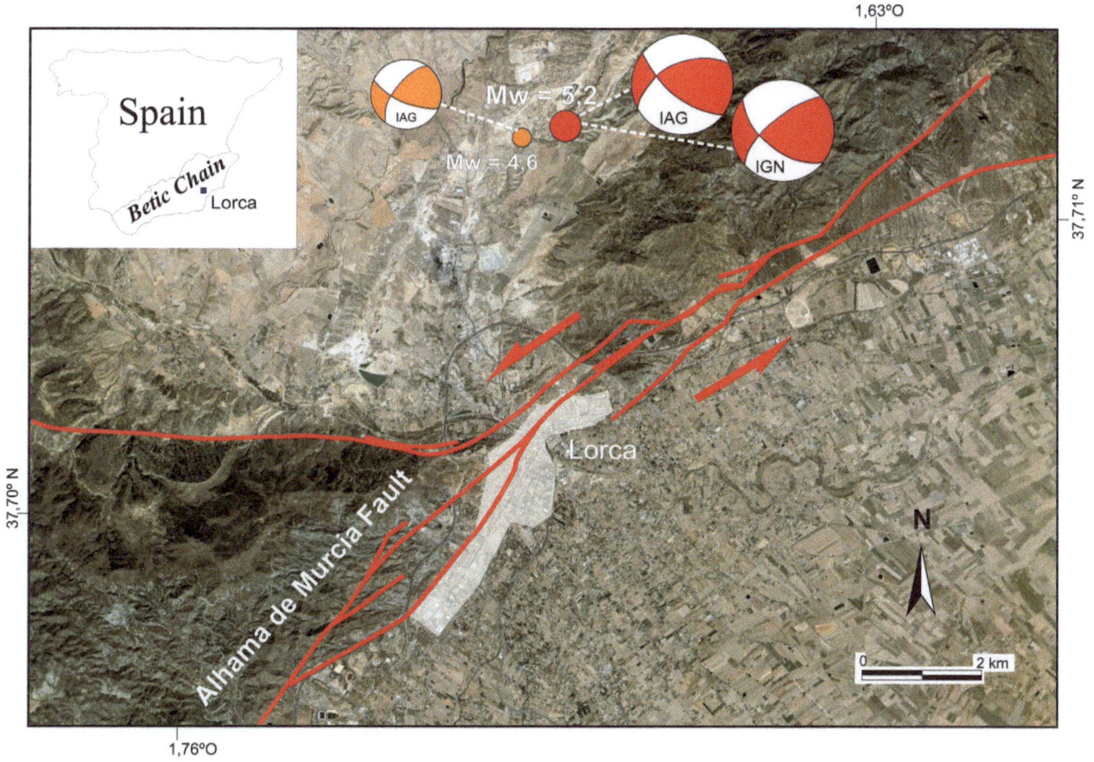

Fig. 5.2 The geographic location of Lorca (SW of Spain) and the Alhama de Murcia Fault (FAM). The red dots are the epicentres of the two earthquakes of the Lorca and the focal mechanisms calculated by different agencies: Instituto Geográfico Nacional (IGN, 2012) and Instituto Andaluz de Geofísica (IAG) (López-Comino et al. 2012)

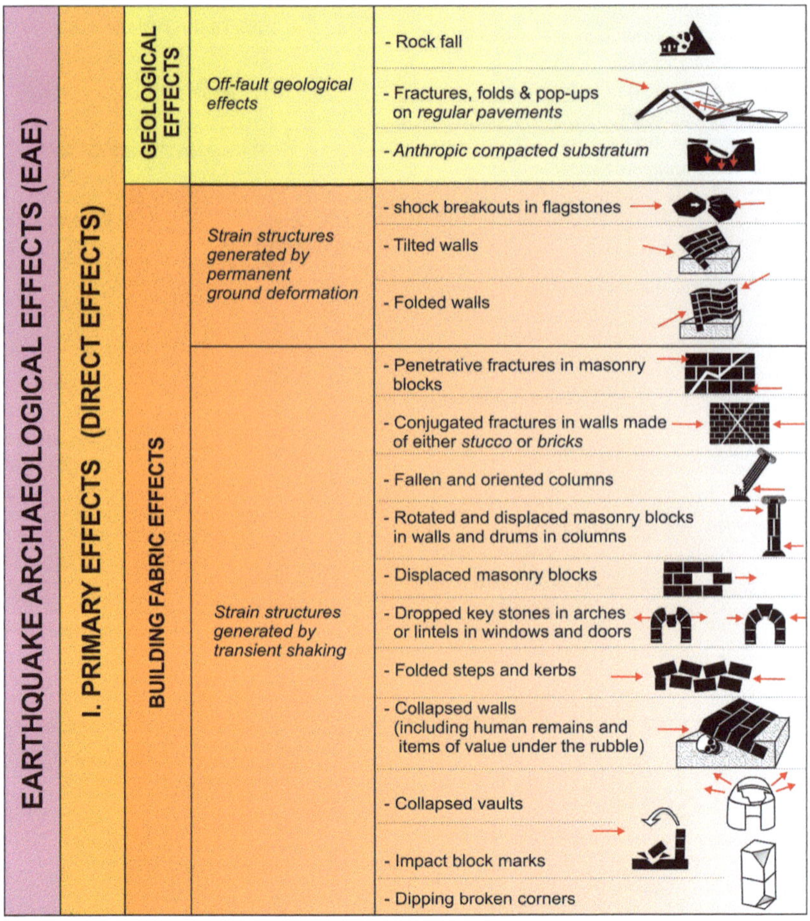

Fig. 5.3 Earthquake Archaeological Effects (EAEs), a classification chart of a compilation of the Cultural Heritage damaged by the earthquake of Lorca. After Rodríguez-Pascua et al. (2011)

sight (IPCE, 2012). In this work, we have carried out a data collection of EAEs recognized in the cultural heritage from the classification proposed by Rodríguez-Pascua et al. (2011) (Fig. 5.3).

5.3 Calculation of the Ground Direction of Movement

Rodríguez-Pascua et al. (2012) collected 127 EAEs in 64 buildings within the historical downtown of Lorca. This part of the city is no regular urban layout of streets, there is no preferential orientation of the streets arranged around the main street. This urban planning was common during the Islamic period of the city (from the eighth to the thirteenth centuries CE), and also common throughout the Islamic period of the Iberian Peninsula (from the eighth to the fifteenth centuries CE). As an advantage, measures of damage obtained from buildings will not be constrained by an orthogonal street arrangement and, therefore, the anisotropy of the seismic ground motion could be better determined.

The classification of EAEs is based on the seismic effects produced during the earthquake, which are considered as (a) coseismic or primary effects (generated during the passing of the seismic waves) and (b) post-earthquake or post-seismic effects (recognized after the seismic ground motion). In the case of the Lorca earthquake, the EAEs that were mapped correspond to coseismic effects (Fig. 5.3), among which we can differentiate two different effects: (1) effects associated with geological secondary effects, such as rock fall on buildings and (2) effects associated with the fabric of the buildings (such as foundation, façade, masonry, concrete covering, and bricks). On the other hand, it is possible to differentiate between EAEs, which are related to the permanent deformation of the substrate and EAEs produced by the transient ground deformation when seismic waves pass through (see Fig. 5.3).

From the 27 different cosmic effects proposed by the classification of Rodríguez-Pascua et al. (2011), a total of 21 types of EAEs were observed and mapped. Below is a summary of these EAEs, showing the more representative examples in Lorca's historical heritage.

5.3.1 Geological Effects

1. *Rock fall*: This phenomenon has been observed mainly associated with the cliffs of the mediaeval Castle of Lorca, located on top of a calcarenitic butte in the NE section of the city. Numerous rock falls of metric size rock-blocks affected the nearby southward houses at the foot of the butte and roads on the north side. It was possible to document both the direct impact of the blocks on the buildings and the impact on the pavement, which generated impact craters and associated cracks (Alfaro et al. 2012).
2. *Fractures, folds, & pop-ups on regular pavements*: These EAEs have been found in different locations in the city of Lorca, a good example being those found on the pavements around the street named "*Tercera Edad*" (Fig. 5.4a) and the Convent of the Clarisas (nineteenth century CE). Figure 5.4a shows a fold-axis with N050°E trend, indicating a shortening direction of the paving with N140°E trend. However, this folding is constrained by the inherent anisotropy from the regular layout of the tiles (orthogonal). Consequently, it is recommended to get a set of these folds in pavements with different orientation to calculate the ground shaking trending.
3. *Anthropic Compacted Substratum*: The compaction of anthropogenic substratum is produced by the reorganization of the classical fabric of grains without the presence of water. This effect generates fractures in compacted sand surfaces and folds in paved surfaces. In this work, a maximum depth of 18 cm of soil subsidence has been measured. As mentioned, the orientations of these structures are determined by the previous orientations of the anthropic constructions. They therefore cannot be used for the calculation of the mean directions of ground movement as isolated data.

5.3.2 Building Fabric Effects

Two types of EAEs associated with deformations generated by "*permanent ground deformation*" are described below.

1. *Impact block marks*: Impact marks when the tiles collide with each other during the horizontal seismic vibration. Figure 5.4b shows an impact between tiles by marking a direction of impact according to the N175°E. Inter-tile impacts are a good kinematic indicator since they may appear both on the edges of the tile and in the corners, thus diminishing the prior conditioning factor that the anisotropy of the regular paving pattern involves.
2. *Tilted and folded walls*: A good example can be found in the remains of the old "*Puente del Oro*", where there is a tilting of a wall panel striking N220°E. The axis of the folded wall tends towards the vertical value, around 85°. This provokes the cracking at the joint with the oblique wall, reaching an opening of 12 cm at the highest part.

The following group of EAEs corresponds to associated effects *generated by transient shaking* of the substrate that affects the building fabric.

1. *Penetrative fractures in masonry blocks*: This strain structure does not indicate directions of movement but can be used in combination with tilted and folded walls and movements in masonry blocks. This structure has been found in all the important masonry buildings,

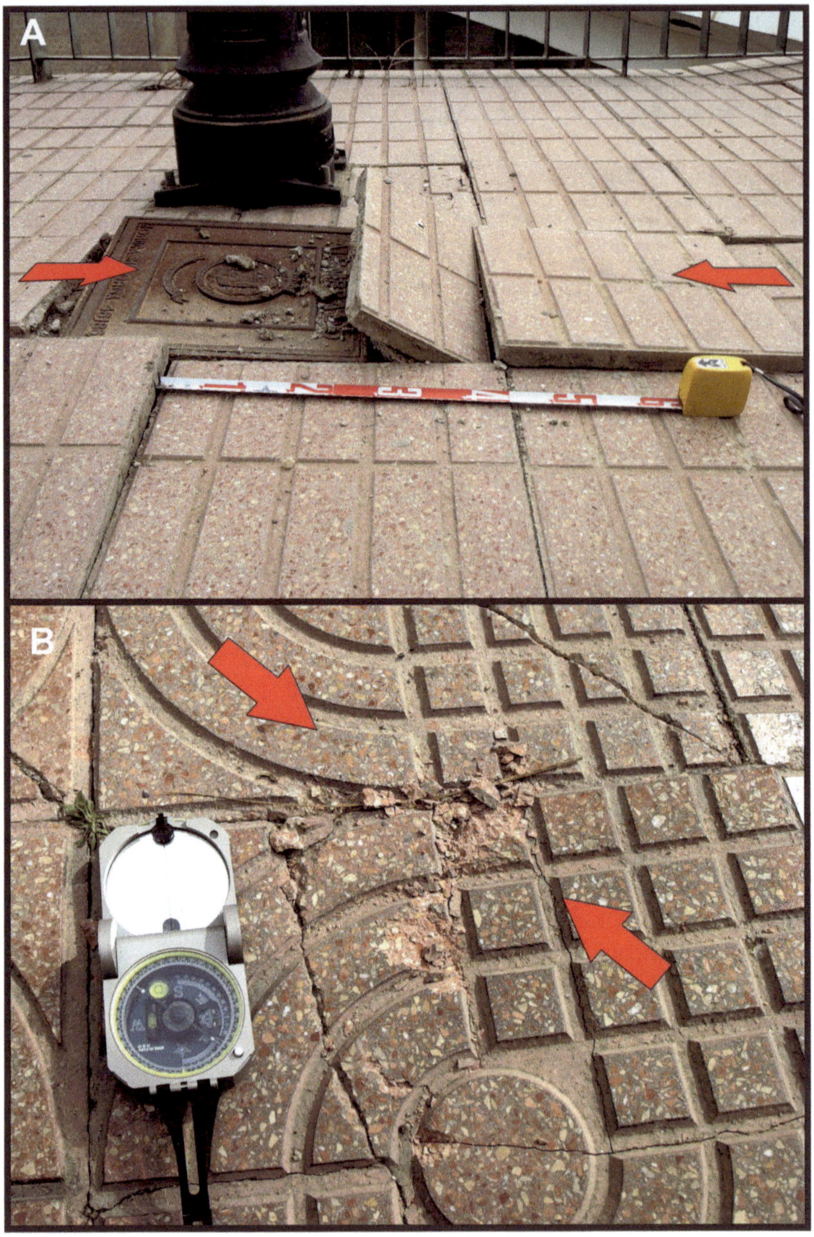

Fig. 5.4 (**a**) Fold in regular flagstones deformed by a shortening direction N140° E. (**b**) Shock breakouts in flagstones caused by a shortening direction N175° E. Red arrows mark the shortening direction

such as the Collegiate Church (from the fifteenth to the sixteenth centuries CE) of *San Patricio* and the sixteenth-century churches of *Santiago, Paso Azul, Paso Blanco, Carmen*, etc. Moreover, these fractures have also been observed in modern buildings such as the built masonry block doorway of the today demolished *Ramón Arcas Meca, a* secondary school, of which only this doorway has been preserved.

2. *Conjugated fractures in walls made of either stucco or bricks*: This is one of the most reliable seismic markers, indicating the direction of the shear stress that generates this type of fracture set. This is one of the most abundant EAEs observed in both historical and

modern buildings. One of the best examples can be seen at the Convent of the Clarisas (nineteenth century CE), which was severely damaged by the earthquake, and it was partially demolished and finally reconstructed. Figure 5.5a shows the genesis of these fractures in the façade wall of the convent's living quarters, striking N150°E. Figure 5.5b shows the same fracture set in the faces of the convent's orchard and with horizontal fractures in the perpendicular wall. The alternating oscillation of the building indicates that the perpendicular walls develop horizontal fractures. Figure 5.5c shows an interpretative diagram of the oscillating movement that is responsible for generating these fractures.

3. *Fallen and oriented columns*: As a general rule, the oscillating movement of columns could tilt the cylinder to any direction, and consequently, columns are excellent indicators of directional movement by ground shaking. Figure 5.6 shows a set of fallen columns with an N330°E trend (NW). In this case, they have a limitation prior to movement, the balustrade that supports it at the top. The columns of strips of the balustrade perpendicular from this direction suffered folds but did not collapse, being compatible with the direction of movement of the NW-SE substrate. Other markers that indicate fall direction are statues, pinnacles, and decorative elements located on the cornices of the buildings. The best example can be found in the pinnacles of the bell tower of the Collegiate Church of *San Patricio*, which were projected towards the N340°E, which broke roofs and left impact marks inside the building (De la Hoz 2012).
4. *Rotated and displaced masonry blocks in walls and drums in columns*: The rotation in the drums, monoliths, statues, and similar architectural elements is the result of a vertical movement that generates the jumping and a repositioning in the fall. Hence, they cannot be used as markers of horizontal movements, although they are excellent witnesses of past seismic activity since these movements can only be generated by seismic shaking and are very difficult to repair. They have been found in the pinnacles of the Collegiate Church of *San Patricio* and the Church of *San Mateo*, showing opposite rotation directions, counter- and clockwise, respectively.
5. *Displaced masonry blocks*: This EAE is a good marker, especially when the blocks of masonry are located at the corners of the buildings, where the degrees of freedom for horizontal movement increase. This displacement indicates the slip-vector (direction, sense, and magnitude) observed in the *Torre del Espolón* in Castillo de Lorca (Fig. 5.7a). The displacement at the base of the column was 16 cm towards the N235°E striking (Fig. 5.7a). On the other hand, the Cemetery of San Clemente, located 4 km southward away of the downtown of Lorca, exhibits an excellent compilation of EAEs affecting pantheons, gravestones, and sepulchres. In particular, displacements of masonry blocks with N125°E trend have been observed (Fig. 5.7b), and with a maximum horizontal displacement of 5 cm. The fall of pinnacles and decorative elements towards N300°E was also documented, with impact marks on the paving that limit some of these pantheons.
6. *Dropped key stones in arches or lintels in windows and doors*: One of the world-wide observed EAE associated with earthquakes is the fall of the key stones in arches. Throughout history, this arch style in architecture is the most used structure to put windows, lintels, and doors in buildings and entrances. The forces which support the arches fall on the sidewall and when the bow starts to oscillate during the ground shaking, the keystones fall and move downwards. However, the upward extrusion of these keystones is the opposite case, and it was not documented until the 2011 Lorca earthquake by Rodríguez-Pascua et al. (2012). The bell tower of the Church of *El Paso Azul* shows a nice example (Fig. 5.8). In this tower the

Fig. 5.5 Conjugated fracture sets in walls oriented N150°E: (**a**) the Clarisas Monastery, (**b**) modern apartment building (sector of the Guadalentin River), and (**c**) sketch of the conjugated fractures generation due to superficial seismic ground oscillation

Fig. 5.6 Fallen and oriented columns of the ancient Infantry Headquarter's balustrade in Lorca: (**a**) The direction and sense of fall was N330°E. The red arrow indicates the focus and purpose of the ground movement responsible for the collapse and (**b**) folded balustrade perpendicular to the collapsed column shown in (**a**)

keystones fell in the arches located in the southeast sidewalls. Furthermore, the orb at the top of the tower also fell, with N340°E trend (Fig. 5.8a). A special case has been documented in the consolidated ruin of the Church of *Santa María* (Fig. 5.8b,c). In this case, the oscillation on the surface with orientation parallel to the direction of the arch (N155°E) produces a horizontal compression of the starting point of the arch over the keystone. Once again, the mean ground direction of movement is consistent with the NW-SE trend.

7. *Folded steps and kerbs*: Few examples were reported at Lorca, being associated fundamentally with the compaction of anthropic substrates, instead of compressions produced by the horizontal ground movement. Folded and fractured kerbs were observed on the pavement of the façade of the Convent of

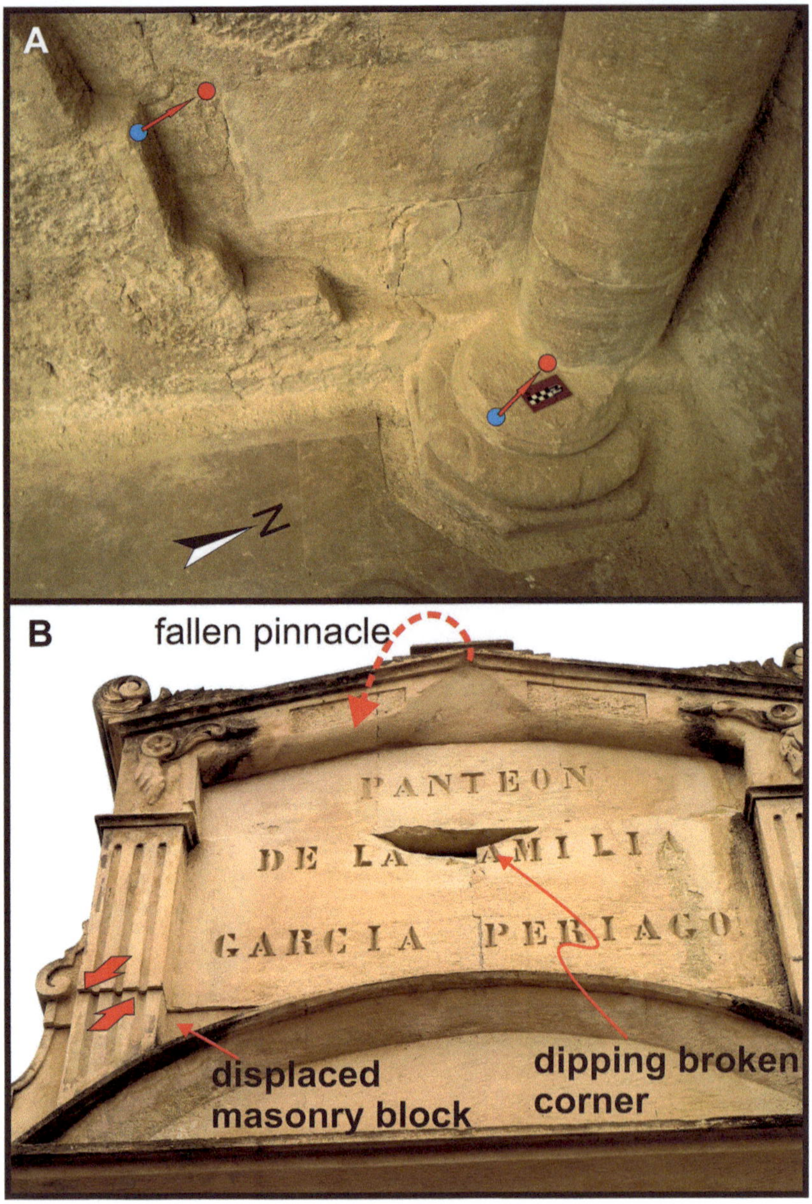

Fig. 5.7 Displaced masonry blocks: (**a**) the red arrow is the vector of ground displacement (16 cm towards N325°E) in the upper room of the Espolon Tower (Lorca Castle). (**b**) Damaged pantheon façade in the San Clemente Cemetery, the direction of red arrows is N125°E. The fall direction of the pinnacles is towards NW

Las Clarisas and in the surroundings of the streets parallel to the Guadalentín River.

8. *Collapsed walls*: Principally walls collapsed in a NW-SE direction. In those cases where the walls are not connected to lateral structures collapses could be observed with direction and sense of movement, fundamentally towards the NW direction.
9. *Collapsed vaults*: The vaults are structures very vulnerable to earthquake shaking, because as they are circular-based structures that are strongly affected by horizontal vibrations. They are excellent earthquake markers, although cannot be used as kinematic indicators, since circular collapses would not indicate any directionality. The vault of the *Santiago* Church was totally collapsed during the earthquake. Nevertheless, the case of the vault of the *El Paso Blanco* Church exhibits an excellent example of the

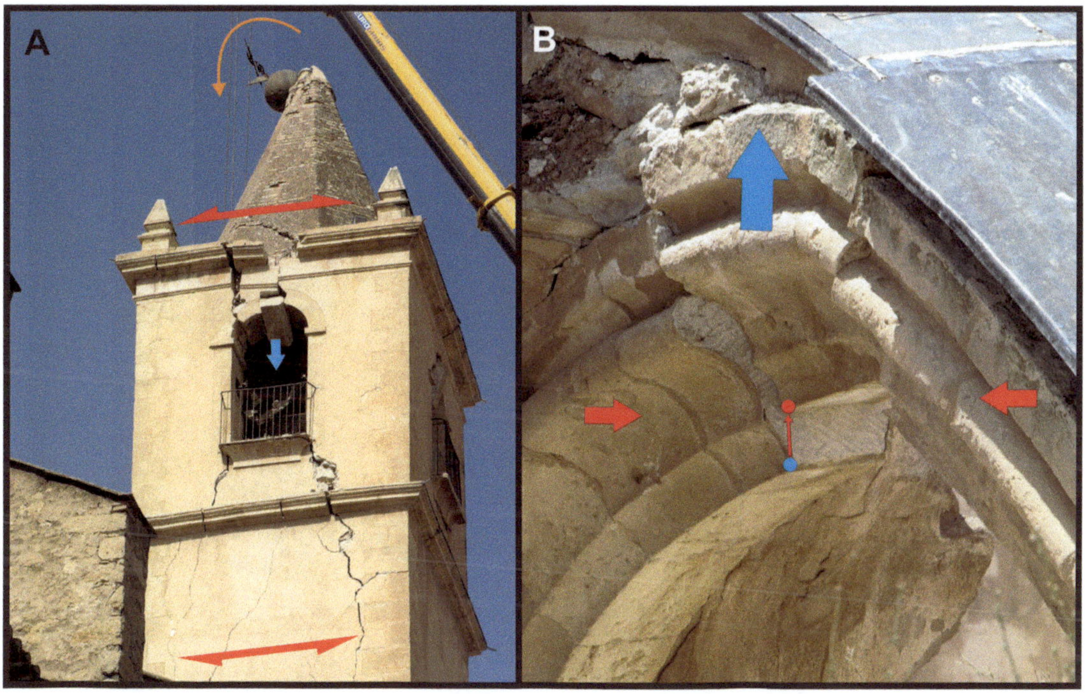

Fig. 5.8 Displaced keystones at (**a**) bell tower of the Paso Azul Church, the direction of the arch is N095°E, note the fall of the final shot of the tower to the west, and (**b**) pop-up of the keystones generated by cyclic shear stress without vertical load of the roof (the direction of the arch is N155°E). Consolidated ruins of the Santa Maria Church

seismic deformation. Restoration works performed on this vault during the 1970s added a 20 cm concrete cap on top of the vault, increasing the weight. A 14 cm horizontal displacement can be observed in the upper cap of the vault with N330°E trend (Fig. 5.9). The fracture and displacement of the vault occurred because of the concrete part of the dome remained static, whereas the base was displaced by the horizontal movement of the nave, generating a fracture at the limit of the concrete area, and remaining intact upwards.

10. *Dipping broken corners*: These EAEs are generated in masonry blocks that oscillate with the cyclic horizontal shear movements. These oscillations produce the collision between blocks fracturing in the weakest areas, the corners. Therefore, the poles of the fracture planes will indicate the direction of mean ground movement. Some outstanding examples of these EAEs can be seen in the pinnacles of the Collegiate Church of *San Patricio* and in the Church of *San Mateo*, as well as inside the Church of *Santiago* and in the other buildings with masonry blocks. The average orientation of land movement that have been obtained is NW-SE.

5.4 Mean Ground Direction of Movement Generated During the 2011 Lorca Earthquake

To calculate the mean ground direction of movement caused by the 2011 Lorca earthquake, the EAEs were all mapped and indicate the direction of movement. These results are conditioned by previous anisotropies such as the orientation of walls. The most common EAEs measured are: firstly conjugated fractures in walls, then horizontal fractures, tilted and collapsed walls, dropped keystones in arches, dipping broken corners,

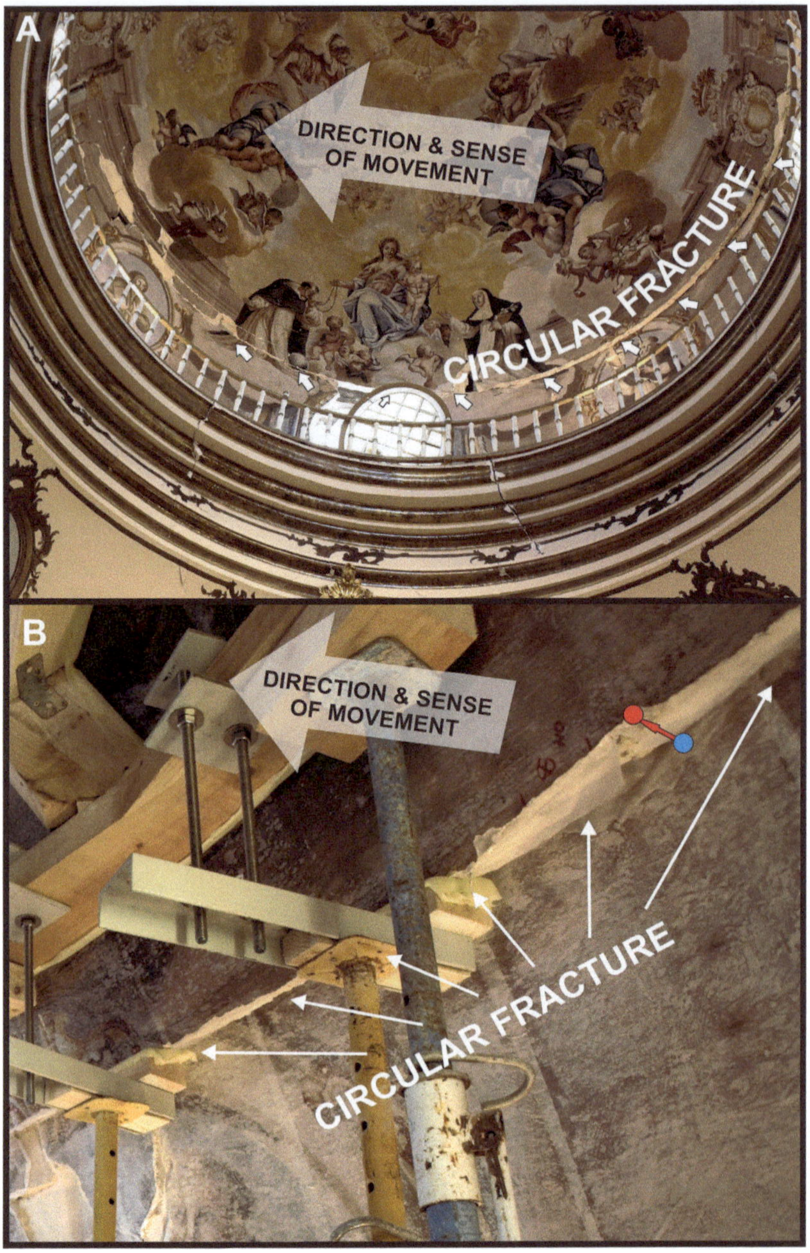

Fig. 5.9 Horizontal fracture of the Church of the Paso Azul dome on the edge of the concrete deck added in recent restorations. (**a**) The white arrow indicates the direction of movement of the dome towards N330°E. (**b**) The dome has a close view of the displacement vector (14 cm in magnitude forward N330°E)

displaced masonry blocks, impacts and finally, folds in pavements and oriented falls from corners.

A total of 127 EAEs were mapped and the mean seismic ground direction of movement is N150°E (Fig. 5.10). This information is consistent with data taken outside the historic quarter in buildings from the twentieth century affected by the earthquake as well as data taken outside the city in the cemeteries of *San Clemente* and *San Cristóbal* (Giner-Robles et al. 2012). The rose diagram of the movement of a particle during the earthquake, extracted from the accelerometer located by the IGN (Cabañas et al. 2014) in the basement of the old prison of Lorca, shows the same pattern of movement as that calculated by EAEs (Fig. 5.10).

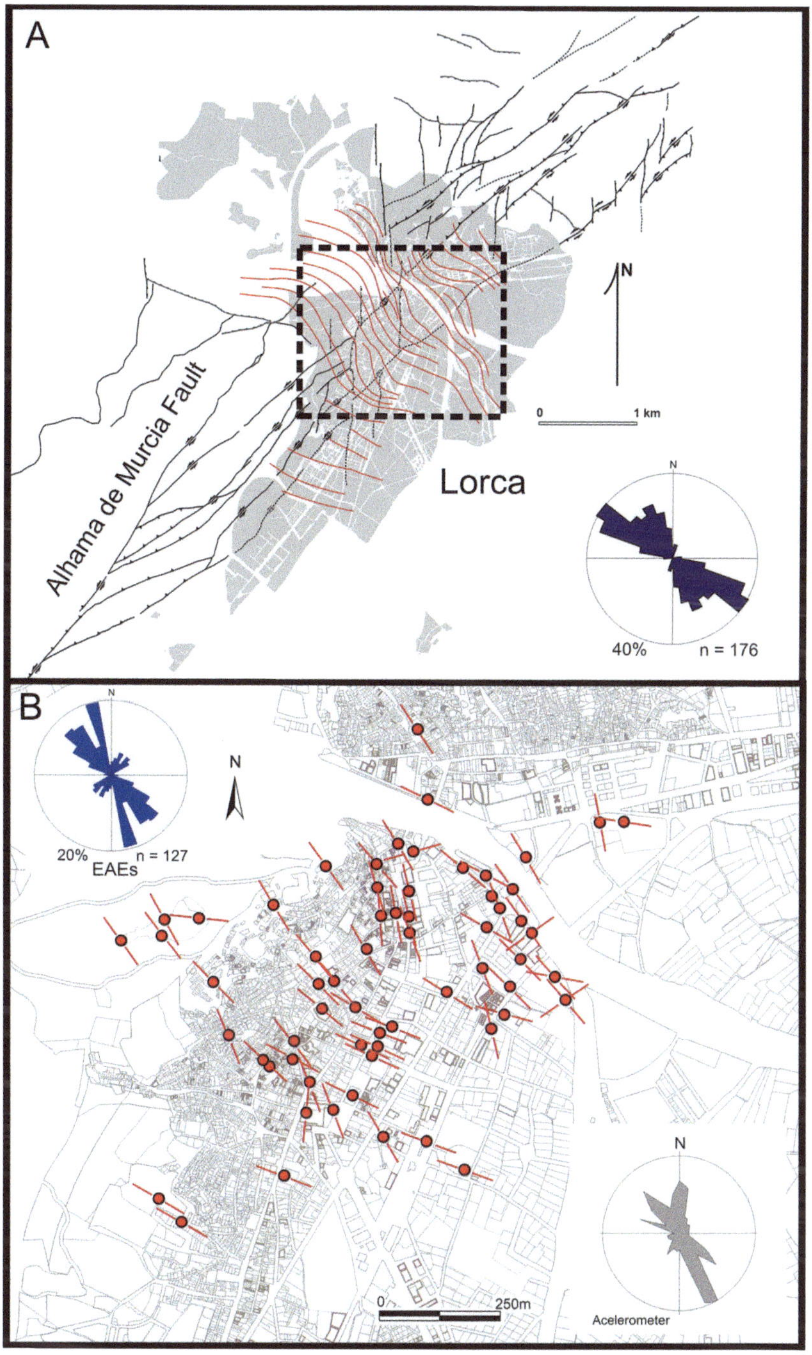

Fig. 5.10 Results of the structural geological analysis of the EAEs in Lorca: (**a**) general map of the Lorca city with the complete results, including cultural heritage and modern residential building, the red lines represent the trajectories of mean seismic ground movement (modified from Giner-Robles et al. 2012) and (**b**) Map of the downtown of Lorca, where they have located the 64 buildings with EAEs (red dots) used for the calculation of the substrate movement direction. Red lines mark the mean directions of movement for each building. The rose of directions shows total values for the directions of motion calculated for the EAEs (upper left) and the directions of motion of a theoretical particle removed from the accelerometer located in Lorca by IGN (modified after Alfaro et al. 2012; Cabañas et al. 2014)

5.5 Conclusions

The results of the mean seismic ground horizontal movement orientations obtained from the mapping of the EAEs, which were generated by the 2011 Lorca earthquake, suggest an NW-SE trend, which agrees with the instrumental data recorded by an accelerometer (Cabañas et al. 2014). The ground seismic direction of movement is perpendicular to the Alhama de Murcia

Fault trace (NE-SW). This result validates the methodology proposed by Rodríguez-Pascua et al. (2011, 2012) and Giner-Robles et al. (2009, 2011, 2012, 2018, 2022), where the building seismic damage generated by an earthquake can be identified, measured, and quantified. This method discriminates the trigger mechanism that generated the deformations in buildings (such as earthquakes, ruins, and wars) since those with a seismic origin are oriented according to the direction of the travelling of the seismic waves throughout the ground. Therefore, the study of EAEs combined with mapping of the oriented damage and active faults is a valuable tool for studying the seismic sources affecting archaeological sites, and a valid indicator for restoration and prevention of future seismic effects in historic buildings. This kind of data and methodologies could be integrated into the seismic preventive national plans for cultural heritage. Finally, the presence of EAEs in archaeological remains is an unequivocal fingerprint of the occurrence of past destructive earthquakes, writing the history of the site and giving an opportunity to increase our knowledge about ancient civilizations.

Acknowledgments This work is supported by the Spanish research project MINECO-FEDER PID2021-123510OB-I00 (QTECTSPAIN). This is a contribution of the QTECT-AEQUA working group.

References

Alfaro P, Delgado J, García-Tortosa FJ, Lenti L, López JA, López-Casado C, Martino S (2012) Widespread landslides induced by the Mw 5.1 earthquake of 11 may 2011 in Lorca, SE Spain. Eng Geol 137–138: 40–52

Cabañas L, Alcalde JM, Carreño E (2014) The observed strong-motion accelerograms from the 2011 Lorca (Spain) earthquake. Bull Earthq Eng 12:1909–1932. https://doi.org/10.1007/s10518-013-9501-0

De la Hoz JD (2012) Terremoto en Lorca: consecuencias y actuaciones sobre el patrimonio religioso. Patrimonio Cultural de España 6:107–122

Giner-Robles JL, Rodríguez-Pascua MA, Pérez-López R, Silva PG, Bardají T, Grützner C, Reicherter K (2009) Structural analysis of earthquake archaeological effects (EAE): Baelo Claudia examples (Cádiz, South Spain). Instituto Geológico y Minero de España, Madrid, p 130

Giner-Robles JL, Silva Barroso PG, Pérez-López R, Rodríguez-Pascua MA, Bardají Azcárate T, Garduño-Monroy VH, and Lario Gómez J (2011) Evaluación del daño sísmico en edificios históricos y yacimientos arqueológicos. Aplicación al estudio del riesgo sísmico. Proyecto EDASI. Serie Investigación. Fundación MAPFRE, Madrid, 96p

Giner-Robles JL, Pérez-López R, Silva PG, Rodríguez Pascua MA, Martín González F, Cabañas L (2012) Análisis estructural de daños orientados en el terremoto de Lorca del 11 de mayo de 2011. Aplicaciones en arqueosismología Boletín Geológico y Minero 123(4): 503–513

Giner-Robles JL, Rodríguez-Pascua MA, Silva PG, Pérez-López R (2018) Efectos sísmicos en yacimientos arqueológicos: catalogación y cuantificación arqueosismológica. Bol Geol Min 129(1/2):451–467

Giner-Robles JL, Rodríguez-Pascua MA, Pérez-López R, Silva PG, Bardají T, Roquero E, Elez J, Perucha MA (2022) Methodological approach for geological structural analysis applied to archaeoseismology. In: D'Amico S, Venuti V (eds) Handbook of cultural heritage. Springer, Cham

IGN (2012) Instituto Geográfico Nacional. http://www.ign.es

IPCE (Instituto del Patrimonio Cultural de España) (2012) Patrimonio en Riesgo: seismos y bienes culturales. Patrimonio Cultural de España, 6: 265p

López-Comino JA, Mancilla F, Morales J, Stich D (2012) Rupture directivity of 2011, Mw 5.2 Lorca earthquake (Spain). Geophys Res Lett 39:L03301. https://doi.org/10.1029/2011GL050498

Martínez-Díaz JJ, Rodríguez-Pascua MA, Pérez-López R, García Mayordomo J, Giner Robles JL, Martín-González F, Rodríguez Peces M, Álvarez Gómez JA, and Insua Arévalo JM (2011) Preliminary geological report of the Lorca earthquake, May, 11th of 2011, mbLg 5.1. Unpublished Report: Instituto Geológico y Minero de España, Madrid, 35p

Martínez-Díaz JJ, Bejar-Pizarro M, Álvarez-Gómez JA, Mancilla FL, Stich D, Herrera G, Morales J (2012) Tectonic and seismic implications of an intersegment rupture. The damaging May 11th 2011 Mw 5.2 Lorca, Spain, earthquake. Tectonophysics 546–547:28–37. https://doi.org/10.1016/j.tecto.2012.04.010

Martínez-Solares JM, Mezcua J (2002) Catálogo sísmico de la Península Ibérica (880 a.C.-1900). Instituto Geográfico Nacional. Ministerio de Fomento, Madrid, p 756

Ortuño M, Masana E, García-Meléndez E, Martínez-Díaz JJ, Stêpancikovâ P, Cunha PP, Sohbati R, Canora C, Buylaert JP, Murray AS (2012) An exceptionally long paleoseismic record of a slow-moving fault: the Alhama de Murcia fault (Eastern Betic shear zone, Spain). Geol Soc Am Bull 124(9–10):1474–1494. https://doi.org/10.1130/B30558.1

Pérez-López R, Rodríguez-Pascua MA, Martínez-Díaz JJ, Martín-González F (2012a) Recurrencia sísmica del segmento Lorca-Totana de la FAM mediante liquenometría aplicada en caídas de bloques al NE de Lorca: los terremotos de 1579 AD, 1674AD y 2011. Geo-Temas 13:1–4

Pérez-López R, Martín-González F, Martínez-Díaz JJ, Rodríguez-Pascua MA (2012b) Datación mediante liquenometría de la caída de bloques y desprendimientos rocosos asociados a la sismicidad histórica en Lorca (Murcia, SE de España). Bol Geol Min 123(4):473–485

Rodríguez-Pascua MA, Pérez-López R, Silva PG, Giner-Robles JL, Garduño-Monroy VH, Reicherter K (2011) A comprehensive classification of earthquake archaeological effects (EAE) for archaeoseismology. Quat Int 242:20–30

Rodríguez-Pascua MA, Pérez-López R, Martín-González F, Giner-Robles JL, Silva PG (2012) Efectos arquitectónicos del terremoto de Lorca del 11 de mayo de 2011. Neoformación y reactivación de efectos en su Patrimonio Cultural. Boletín Geológico y Minero de España 123(4):487–502

Silva PG, Borja F, Zazo C, Goy JL, Bardají T, De Luque L, Lario J, Dabrio CJ (2005) Archaeoseismic record at the ancient Roman City of Baelo Claudia (Cádiz, South Spain). Tectonophysics 408(1–4): 129–146

Silva PG, Reicherter K, Grützner C, Bardají T, Lario J, Goy JL, Zazo C, Becker-Heidmann P (2009) Surface and subsurface paleoseismic records at the ancient Roman city of Baelo Claudia and the Bolonia Bay area, Cádiz (South Spain). In: Reicherter K, Michetti AM, Silva PG (eds) Paleoseismology: historical and prehistorical records of earthquake ground effects for seismic hazard assessment. Geological Society of London, Special Publication, vol 316, pp 93–121

Silva PG, Giner-Robles JL, Reicherter K, Rodríguez-Pascua MA, Grützner C, García Jiménez I, Carrasco García P, Bardají T, Santos G, Roquero E, Röth J, Perucha MA, Pérez-López R, Fernández Macarro B, Martínez-Graña A, Goy JL, Zazo C (2016) Los terremotos antiguos del conjunto arqueológico romano de Baelo Claudia (Cádiz, Sur de España): Quince años de investigación arqueosismológica. Estud Geol 72(1): 1–24

Part III

Anti-seismic Measures in Ancient Building Construction and in the Vernacular Tradition

Anti-Seismic Measures in the Building Tradition

6

Laura Pecchioli and Giovanni Cangi

Abstract

The issue of historic and pre-modern anti-seismic techniques concerns many contexts. The logic of construction and consolidation is primarily based on intuition and experience. It concerns valuable building typology, particularly ordinary buildings, on which techniques are more frequently experimented with, refined, and developed to solve complex problems in simple methods.

In this way, construction types became popular because they met functional requirements and were reliable and safe. The structural measure standard of the load-bearing walls and the dimensions of the masonry frameworks vary. They vary in relation to the resistance characteristics of the beams and vaults, elements that characterize all historical buildings and define the typical contours of a traditional building designed to withstand seismic action. Thus, there is a direct connection between intuition and engineering based on the science of construction, which must be recovered.

The most common seismic prevention techniques, such as spurs, tie rods, and buttressing arches, can demonstrate it. These elements were built based on simple geometric criteria, but today they can be dimensioned and verified in quantitative terms to optimize their costs and function. Among the requirements guiding the construction logic of historic buildings, rule predominates toward the construction of structural units placed side by side and joined to form unitary organisms, where interactions and reciprocal collaboration constitute the strength of the structures. An opposite logic to that pursued in modern constructions is based on creating technical joints between side-by-side structures, which must be kept structurally separated to prevent any interaction. Based on laboratory experiments supported by Avant-garde research, Seismic regulations usually pay little attention to ancient technologies. Urban plans tend to avoid modifications to the consolidated historical building, prohibiting one of the most frequent structural strengthening interventions: added bodies outside the building. In the building tradition, to counteract the outward rotation of the façades, it used to add a new body to the building to increase its

L. Pecchioli (✉)
Institut für Archäologie, Klassische Archäologie, Winckelmann Institut, Ostia Forum Projekt (OFP), Humboldt-Universität zu Berlin, Berlin, Germany

Institut für Kunstgeschichte, Bauforschung und Denkmalpflege, Technische Universität Wien, Vienna, Austria

Dipartimento di Architettura e Restauro, Studi Urbani di Roma, Università di ROMA TRE, Rome, Italy
e-mail: laura.pecchioli@hu-berlin.de

G. Cangi
ITABC (Istituto per le Tecnologie Applicate ai Beni Culturali), CNR (Consiglio Nazionale delle Ricerche), Rome, Italy

L. Pecchioli (ed.), *Archaeoseismology*, Natural Science in Archaeology,
https://doi.org/10.1007/978-3-031-28303-1_6

primary inertia and resistance to horizontal actions or to create arches or vaults through the narrowest alleys to join the building opposite. This way, the structure was strengthened, and the rooms could also be enlarged. Another rule of the building tradition was the gradual reduction of the masses in elevation, through the tapering of the walls and the construction of lightweight floor systems, in many cases challenged by current techniques.

Keywords

Seismic prevention techniques · Damage · Recovering practical techniques · Conservation codes · Seismic vulnerability · Local seismic culture

6.1 Introduction

In the Italian context, restoring historical anti-seismic culture is a priority. Reinforcing the level of maintenance and paying greater attention to the mistakes of the past can provide greater control in preventing seismic damage. However, a more comprehensive approach, including an interdisciplinary approach at the urban level, with interventions designed not on a single building, requires a broader knowledge of the anti-seismic safeguards inherited from the past (Pecchioli et al. 2022). In our architectural heritage, we are lucky enough to identify, in various urban structures, those measures favorable to a reduction in vulnerability, which are often underestimated.

Although historical and other older buildings can be adapted to survive earthquakes, doing so often damages or destroys the features that make such buildings significant. While life safety issues remain paramount, various approaches can help in protecting historic buildings from the devastation caused by earthquakes, as well as the damage inflicted by well-intentioned but unsuitable retrofit procedures (Cangi 2014, 2019). Building owners, managers, consultants, and communities should be actively involved in planning and recovering practical techniques and traditions to deal with these threats. This contribution focuses on the importance of urban form as a factor in these efforts. It aims to briefly introduce some of the methods of seismic strengthening inherited from the past.

6.2 Conservation Codes

Historic preservation ethics guides responsible retrofit decisions and implementing of seismic rehabilitation, but methods based on modern solutions are often suggested without analyzing landmark construction and urban evolution.

Undertaking seismic rehabilitation of a historic building requires careful planning and the collaborative work of architects, engineers, code officials, contractors, and agency administrators. An interdisciplinary team can ensure that historic buildings' architectural, structural, financial, programmatic, cultural, and social values are preserved while making them safe for continued use.

Major repairs and alterations, additions, changes in use, and local ordinances can trigger a review of a structure's compliance with current building code requirements, including seismic strengthening requirements that specify a minimum level of protection from seismic hazards. These may specify the path to be followed for assessing seismic vulnerability, outline rehabilitation methods, establish limits on the choice of structural analysis procedures and acceptable rehabilitation strategies, and designate alternative compliance methods.

Building codes are primarily intended to guide the design and construction of new buildings and often require methods critical to satisfactory seismic performance. However, determining the seismic adequacy of historic buildings by comparing them to the requirements for new construction can be difficult, even impossible, and inappropriate, partly because new building codes do not cover some traditional construction materials and methods.

Many prescriptive code-compliant standards can produce significant alterations to the appearance and integrity of historic buildings. The most cost-effective ways to reinforce a structure include introducing an entirely new structural system, adding new elements, and filling in irregularities or large openings without considering how these modifications affect the architectural design. Unfortunately, the results of these approaches can be quite invasive. Successful seismic rehabilitation of historic buildings requires the skillful use of the best available technology, an adequate understanding of historic preservation, and knowledge of the intrinsic strength of traditional materials and structural systems (Giuliani 2011).

The exciting research carried out by the European Centre for Cultural Heritage in Ravello is part of the Council of Europe's EUR-OPA agreement on major risks in areas at seismic risk. The program aims to identify traditional anti-seismic building techniques for revitalization, drawing on know-how within the local community. This requires analyzing techniques and methods developed over the centuries in local communities. The multidisciplinary research program established for this purpose has adopted the archeological approach, recognizing that traditional physical and material assets provide a fundamental source of information (Helly 1995).

Another contribution that deserves attention is attributed to Nicola Ciaburri (2017), who analyzed urban centers' vulnerability and earthquake culture. The published text focused on a view of the seismic problem on an urban scale, highlighting vernacular solutions as technical garrisons that should be analyzed, preserved, and taken into account in addressing current solutions.

The objectives of successful seismic rehabilitation should be to reduce seismic vulnerability, maintain constructive features and historical value to the greatest extent possible, and avoid or minimize the impact of alterations, especially on the most valuable features (Giuliani 2016). Historic buildings are particularly vulnerable to seismic events, especially those built before seismic codes were implemented. Moreover, more and more communities are adopting higher standards for seismic retrofits on existing buildings. Historical features and materials, both structural and non-structural, should be preserved and maintained, not as museum artifacts, but to continue to perform their historical function as much as possible, rather than being replaced wholesale in the seismic strengthening process.

6.3 Local Seismic Culture

In a given society, the existence of a local building culture implies developing a process of awareness and sedimentation of knowledge about the unspoken rules which define the building systems. A seismic culture can be described as the body of knowledge, both pragmatic and theoretical, accumulated in a community exposed to seismic hazards over time.

Local seismic cultures include seismic norms that have not been formally established in a written code but are nevertheless visible in the characteristics of buildings, site selection, and the general layout of the territory (Ferrigni et al. 2017). These elements have now become standard practice and heritage.

If reconstruction begins after the earthquake, urban plans are often based on linear and reticular schemes, very different from the previous organic systems, such as those of the Middle Ages. The streets are more significant, with wide spaces designed to create safe outdoor areas. The origins and persistence of local seismic culture can be determined both by the scale of intensity and frequency with which earthquakes occur and by economic and social conditions, including the availability of resources and cultural traditions. Ancient builders used well-known construction criteria and devices to build houses capable of resisting earthquakes, refining them with time and experience (Giuliani 2011; Pecchioli et al. 2018). In this way, technological evolution has developed through experimental tests based on direct observation of actual behavior, following telluric forces. Anti-seismic vernacular reinforcements are numerous and, in many cases, depending on available materials, local building cultures, and builders' skills. Recurring defense mechanisms consist of the following:

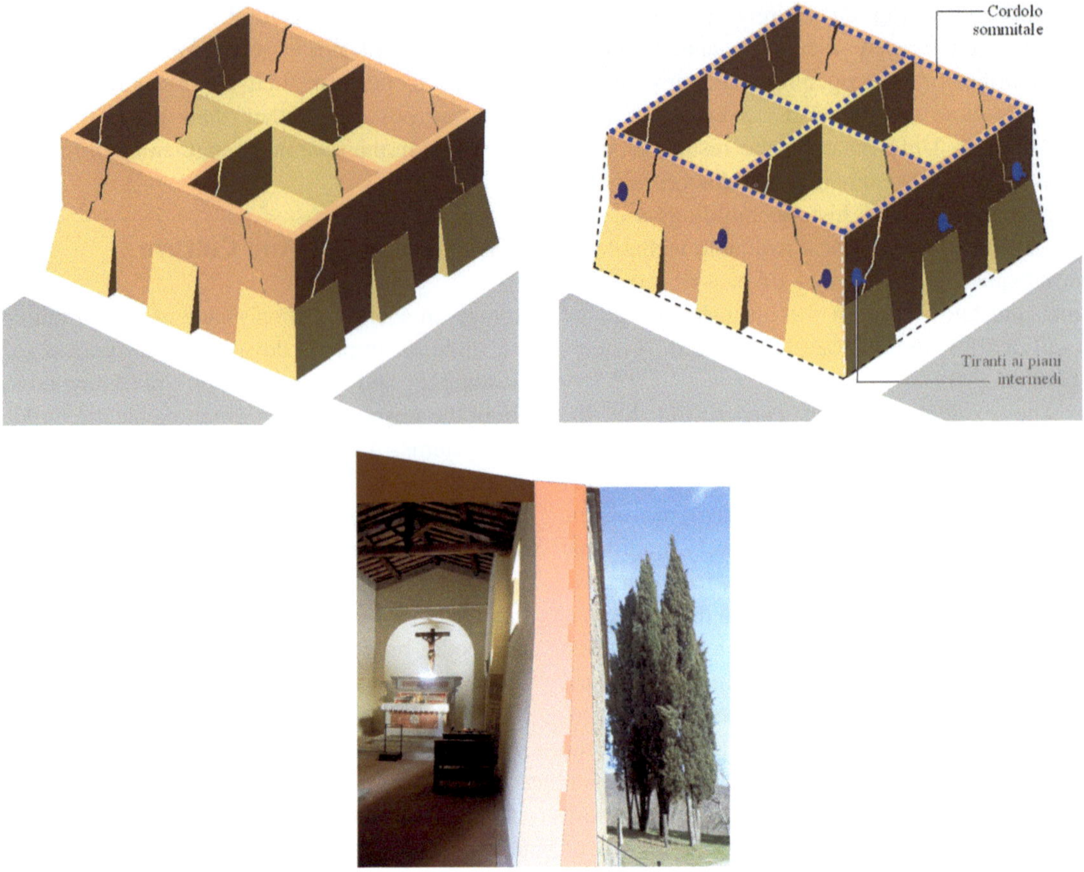

Fig. 6.1 Spurs applied to the outside of damaged buildings may be of limited effectiveness due to the impossibility of extending their base toward the alleys and streets of built-up areas. This condition requires the application of restraining elements at the intermediate and top levels, where the spurs do not produce stabilizing effects

- Connection and restraint systems
- Buttresses
- Large section walls
- Reinforcing arches
- Technical solutions aimed at favoring box-like behavior.

In seismic regions where stone, earth, or brick masonry is the prevalent construction technique, the most frequent prevention and reinforcement measures consist of adopting the mechanism of mutual contrast between buildings or parts of buildings to counteract horizontal forces. Among the traditional devices most commonly used to counteract horizontal forces are: spurs and buttresses, wall supports, stairs, loggias, and contrasting arches, which are often found in correspondence with the walls and cantilevers at floor level (Fig. 6.1).

6.4 Form as Seismic Resistance

In regions affected by frequent and high-intensity earthquakes, local communities have developed strategies to protect themselves from risk (Giovanetti et al. 2018), such as specific building systems or devices designed to reduce the vulnerability of their building habitats (Ciaburri 2017). The extraordinary quality of some traditional building practices and tips, and their relevance over time and during the seismic phenomenon,

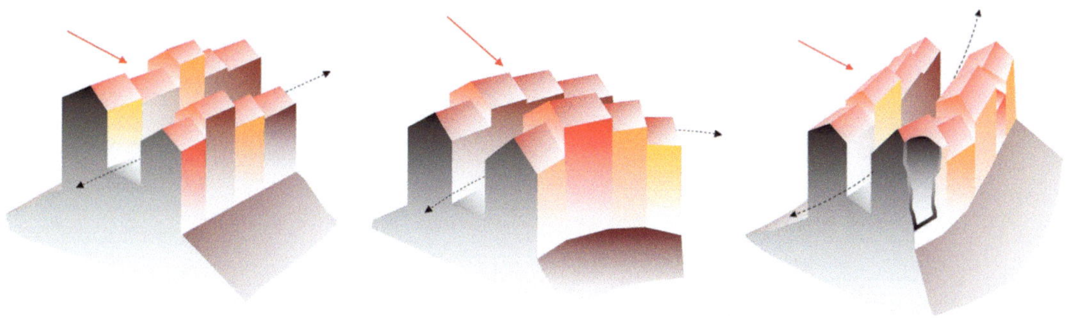

Fig. 6.2 Different effects produced by orthogonal seismic actions acting on linear, concave, and convex ridge building aggregates

lead us to study these techniques to understand them and codify the unspoken rules to regulate the local anti-seismic systems. The various villages developed according to spontaneous building with successive stratifications, which could be articulated or incongruent. The damage caused by recent earthquakes has affected many of our historical centers. The safety assessment and the design of improvements to building units can interact by exchanging mutual stresses, like a domino effect. Therefore, within each building aggregate, it is essential to distinguish the individual structural units, especially if those who study them are called to deal with seismic safety analysis (Fig. 6.2). In this case, it will not be realistic to model the building unit as isolated. Still, instead, the interaction of adjacent structural units must be assessed. The problem remains diversified according to the complexity and articulation of the planimetric and altimetric development of the buildings within the aggregates.

In building a historic city, we face a progressive filling of spaces, adjacent and continually with the existing built-up area, creating the so-called "blocks". Each building aggregate is thus characterized by heterogeneity of wall structures due to a process of "assembly." Recent interventions often see reinforced concrete elements as the protagonists in the realization of *superelevations* or extensions, often structural elements that remain independent from the original. Stiffening interventions are added: structural modifications increase reading difficulty and identify the structural scheme and the "single homogeneous structural units."

The use of a virtual arch model makes it possible to interpret the behavior of block facades (plug, short, or terraced), which have compositions compatible with the kinematic model illustrated above (Fig. 6.3). Furthermore, the case of the head or corner building cell can also be interpreted using this method (Cangi 2005).

6.5 Existing Earthquake Resistances

To guarantee the seismic safety of building aggregates, it is essential to study the role of pre-existing anti-seismic safeguards resulting from previous seismic improvement works and contribute to the aggregate's dynamic behavior. For example, in historical masonry structures with vaults, domes, and high walls, for safety against static loads and dynamic actions such as wind and earthquake, buttresses are among the decisive elements in giving stability to the structures they are leaning against (Pecchioli et al. 2020; Pecchioli 2022). To this end, a contribution by Gürel (2021) considers an actual historical structure in masonry covered by a barrel vault and with a simple and regular shape in plan and elevation. A calculation and base model was created from a section of this structure. For masonry, a macro-modeling approach was used, adopting

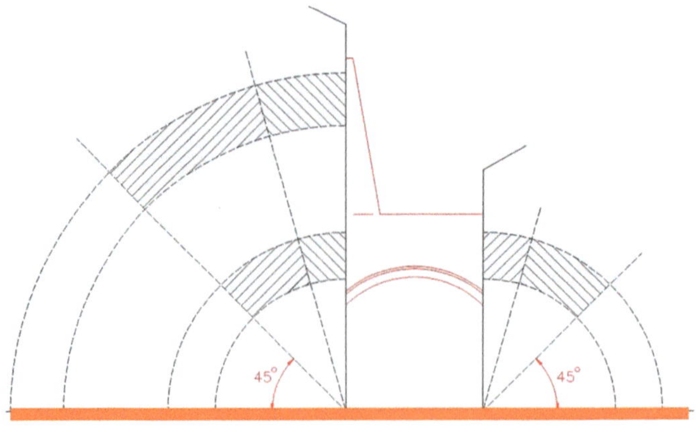

Fig. 6.3 Spur set in elevation on *sbadacchio* (arch. Caprarola, Viterbo)—Historical center. The yawning arches connected between the aggregates to improve their cooperation in an earthquake, limiting damage to the end buildings)

the method of non-linear static analysis and finite element modeling (FEM).

Through these analyses, comparing the base shear forces obtained for the models, the effectiveness of the form of the buttresses on transverse seismic resistance was determined and evaluated.

Among the techniques of seismic prevention, the most frequently used in pre-modern construction is that of the *sbatacchio* arches (or *sbadaccio*), which act similar to the wooden provisional works used to prop up opposing buildings, but which, unlike the latter, are intended to remain as stable anti-seismic protection (Fig. 6.3).

Often the designer is commissioned by a single owner to assess the safety and the resulting consolidation measures for the single US.[1] When analyzing the single US, it is necessary to expand the modeling to include parts of the structure that do not belong to the assessed property. Although with limited information about neighboring properties, it would be relevant to determine how structural continuity with other parts influences the structural response under assessment. This approach translates into evaluating the actions (horizontal and vertical) from adjacent US floors or walls, including the thrusts of arches and vaults, the thrusts of contrasting arches, or tie rods anchored on adjacent parts of the building aggregate (Fig. 6.4).

Another system, in line with the building tradition and among the most effective used to date for controlling damage mechanisms and hypothesizing for walls in an elementary cell, is the insertion of tie rods. This solution performs a prevalent action of contrasting the overturning to which the exposed walls are subject. Tie rods are monolateral restraints, capable of exerting their effort only when subjected to tensile stresses. Usually, in the numerous cases found in historic buildings, the ties are arranged perpendicular to the walls to be connected. However, their diagonal arrangement, which is rarely observed in historical cases, may constitute an improvement. The crossed ties, combined with a system of chains arranged flush with the wall, make it possible to obtain a much more reliable configuration, with the bracing counteracting the planimetric deformations of the floor (Cangi 2012).

In particular, the modeling of the single US must contemplate the "hammering effects" coming from the bordering floors, above all caused by horizons offset in height on the walls in common with the adjacent US. In addition, local effects caused by elevation misalignments, height or stiffness differences between the contiguous US,

[1] Stratigraphic Unit (US) or Masonry Stratigraphic Unit (USM) in other disciplines.

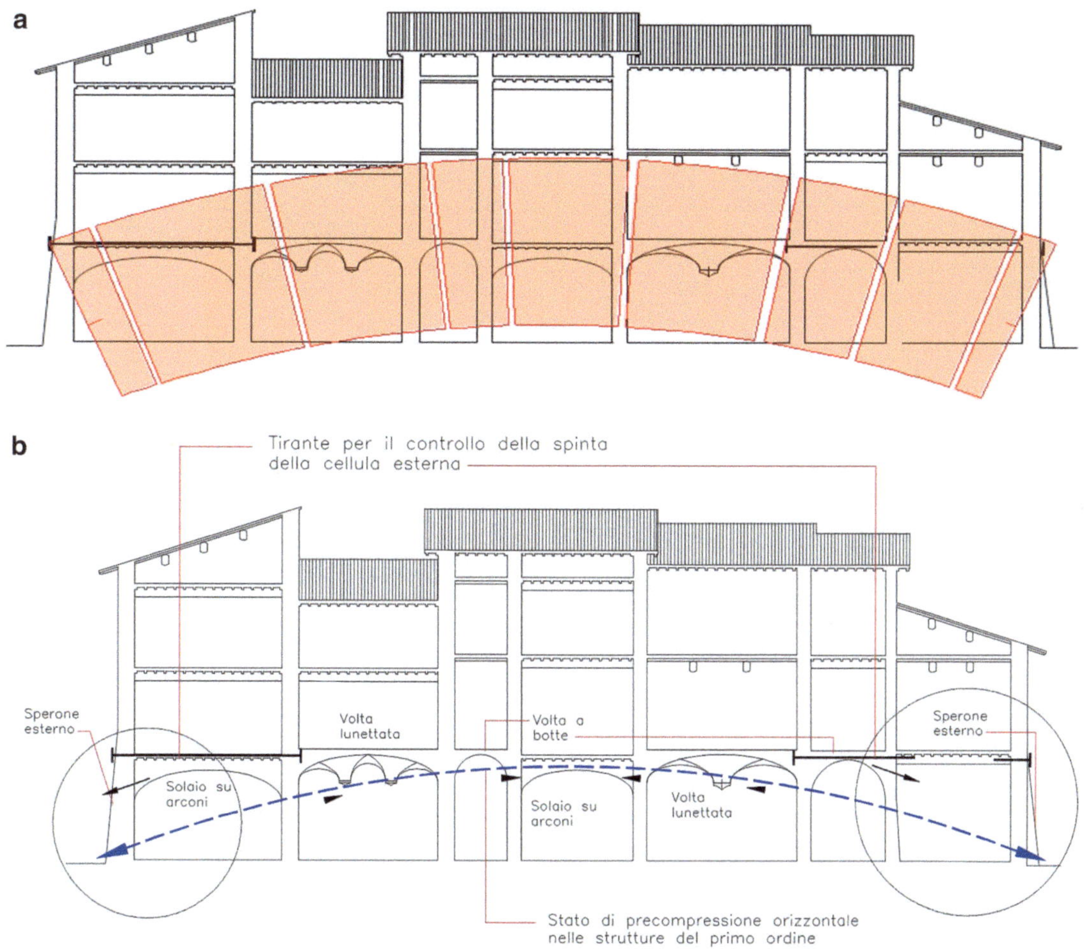

Fig. 6.4 (**a**) Equivalent arc model of a building aggregate: cell of a terraced block with contrasting action provided by adjoining bodies; (**b**) Stabilizing effect produced by the succession of pushing elements in composite blocks of terraced houses

and tilting and translational actions at the end of the US must be evaluated.

The construction of the arches across the streets responds to a simple and intuitive rule, applied precisely in the construction of temporary works, and today unjustly excluded from the list of definitive interventions. "Sbatacchio arches" make it possible to solve static problems that would otherwise be difficult, with a straightforward and effective device against overturning mechanisms. Today, a significant obstacle to applying this technique is the seismic legislation, which does not favor the search for structural collaboration between adjacent buildings and pursues opposite objectives. Another obstacle is that an agreement between the owners of the neighboring buildings is always required to construct arches. The search for cooperation between adjacent buildings, on the other hand, is inherent in the traditional building systems themselves and finds its most significant expression in the building aggregate. The presence of the contrasting arches that survive in historic settlements is not always justified by a real need but sometimes with the opportunity to create functional links between buildings, which over time have become

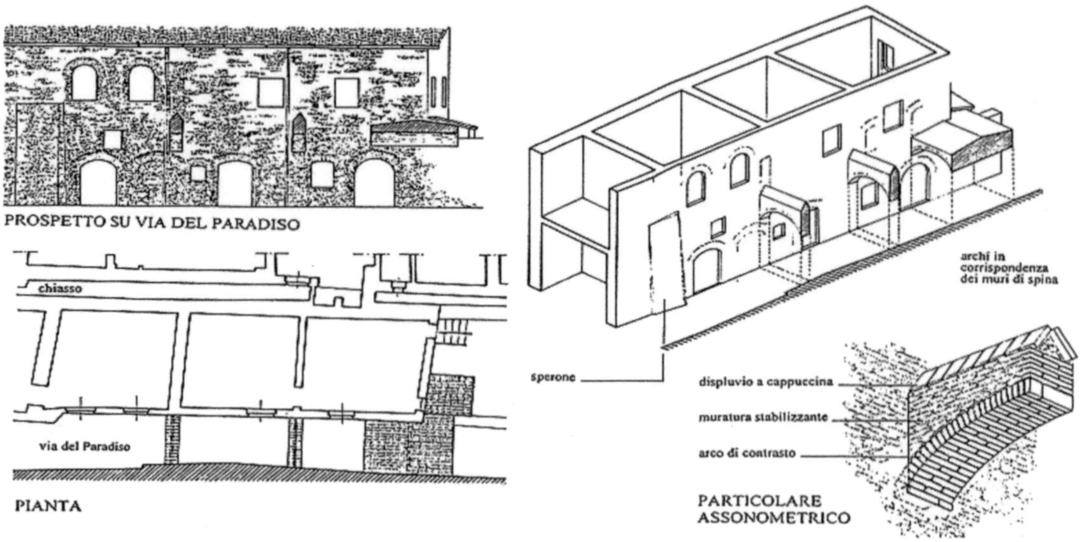

Fig. 6.5 Succession of contrast arcs in serial construction

functional appurtenances over the public street. In some cases, the arches are replaced by vaulted systems supporting the infill above.

The arches' effectiveness depends on the impost's height and the buildings' height. In the blocks of serial houses of the Middle Ages, the arches are generally placed on the first level. At the same time, the contrasting structures in the most important buildings are set in two orders, making them much more effective (Fig. 6.5).

6.6 Conclusions

Unfortunately, anti-seismic solutions can be lost over time, and their reconstruction remains challenging due to the various interventions of complete replacement and consolidation. We speak of collective experience in dealing with what we call "a local seismic culture", in which building rules were considered as models of building systems interpreted as a whole. History shows masonry buildings have resilience; on the contrary, reinforced concrete structures decrease over time after a few years of construction without adequate preventive maintenance. This last aspect is often underestimated. Variables such as shape, materials, and layout in response to seismic stresses influence the structural behavior, and the unknowns relating to the quality of the mortar and building materials always remain. In the case of historical centers in Italy, we can say that most buildings have a masonry structure and are organized in unbroken aggregates. Therefore, it is necessary to evaluate whether "targeted" interventions, in historical urban reality, are planned at different scales: as a single building, an aggregate, and at the city level. In identifying and applying such seismic protection solutions, one can aim to decrease seismic vulnerability and develop an economic and cultural recovery of historic Italian centers.

References

Cangi G (2005) Manuale del Recupero Strutturale e Antisismico–DEI, Tipografia del Genio Civile di Roma, Roma

Cangi G (2012) Manuale del Recupero Strutturale e Antisismico (II edizione), Edizione DEI, Tipografia del Genio Civile di Roma

Cangi G (2014) Tecniche antisismiche nell'antichità, in a. cEntRoni (a cuRa Di), Attualità del- le aree archeologiche: Esperienze e proposte: Atti del VII Convegno Nazionale ARCo 2013, Sezione Protezione dai Rischi Ambientali, Gangemi Editore, Roma

Cangi G (2019) Recupero e Conservazione, «ReC Magazine», Rivista bimestrale (ISSN 2283–7558), Luglio/

Agosto, n. 154, Costruzioni storiche in muratura. Modellazione strutturale schematica

Ciaburri N (2017) La forma come resistenza sismica–una città ricostruita dopo il terremoto del 1688., TETAprint

Ferrigni F, Della Pietra A, Sorrentino MC (2017) Rafforzamento appropriato dell'edificato storico in zona sismica, nell'ambito di LAReHBA Project, Local Appropriate Retrofitting of Historical Built-up Areas, CUEBC

Giovanetti F, Zampilli M (2018) Dopo il terremoto...Come agire? RomeTre-Press, Giornata di lavoro sui recenti eventi sismici

Giuliani CF (2011) *Provvedimenti Antisismici Nell'antichità*. In: «Rivista Di Topografia Antica», 21: Atti Del VII Congresso Di Topografia Antica, Parte III, (2012)

Giuliani CF (2016) Il quadro fessurativo nello studio dei monumenti antichi, «Quaderni di archeologia e di cultura classica», n. 4, Tiburis Artistica Editrice

Gürel M, İzol R, Mollamahmutoğlu Ç, Pekgökgöz R, Yüksel F, Özeren M (2021) Effect of buttress form on transverse seismic resistance of high masonry walls. https://doi.org/10.21203/rs.3.rs-233787/v1

Helly B (1995) Local seismic cultures: a European research program to protect traditional housing stock. Ann Geophys 38. https://doi.org/10.4401/ag-4059

Pecchioli L (2022) Terrae Motus: repair and prevention in Ostia, the harbor city of ancient Rome. In: Compatangelo-Soussignan R, Diosono F, Le Blay F (eds) Living with seismic phenomena in the Mediterranean and beyond between antiquity and the Middle Ages. Archaeopress, printed paperback and electronic book (e-Pdf) as a free download

Pecchioli L, Cangi G (2022) Interpretazione degli effetti sismici eanalisi dinamica dei collassi in Ostia Antica. RA Restauro Archeologico, Florence. https://doi.org/10.13128/rar-11338. Accessed 22 Dec 2021

Pecchioli L, Cangi G, Marra F (2018) Evidence of seismic damages on ancient Roman buildings at Ostia: an arch mechanics approach. J Archaeol Sci Rep 21: 117–127

Pecchioli L, Panzera F, Poggi V (2020) Earthquakes and cultural heritage. J Seismol 2021

The Temple B in Pietrabbondante (Isernia, Italy) as a Case Study of *Seismic Shield* Properties of Roman-Italic Temple Foundations and Podiums

7

Francesca Diosono, Aguinaldo Fraddosio, Alberto La Notte, Nicola Pecere, and Mario Daniele Piccioni

Abstract

The Roman-Italic temples of central Italy, located along the Apennine ridge, an area of high seismic risk, have often been damaged by earthquakes and then reconstructed, as reported even in ancient sources. The Roman religious tradition prescribed that the temples be elevated on a *podium,* that acts as a sacred space, as the foundation, and as an architecturally defined above-the-ground element. Utilizing a multidisciplinary approach that combines a deep archaeological and architectural knowledge of ancient constructions with expertise in seismic engineering and the analysis of mechanical wave propagation, the contribution illustrates the current results on the 'seismic shield' of a selected case study, the Temple B of Pietrabbondante, whose foundation's walls shape cellular compartments filled by layers of clay coming from the surrounding area. A three-dimensional, multiphysics, finite element, numerical model—taking into account either the geometry and the material of the foundations or the mechanical features of the soil between foundation walls—permits the simulation of the propagation of seismic waves towards the construction and the analysis of their impact on the foundation system. Underlying this working hypothesis is a recent field of research concerning 'seismic metamaterials', i.e. the introduction into foundation soils of suitable elements capable of deviating seismic waves, thus protecting constructions from earthquakes. The 'seismic shield' effect mostly depends on the geometric arrangement of the artificial aspects introduced into the soil: this led to thinking that colossal buildings, such as amphitheatres, have survived several seismic events over the centuries due to the particular cellular morphology of the foundations. It is a fascinating hypothesis, opening the way to new interpretations of the capacity of ancient monumental buildings to withstand earthquakes.

Keywords

Seismic shield · Roman-Italic temples · Podium · Seismic metamaterials · Central Italy · Pietrabbondante

The 1st paragraph is by F. Diosono; the 2nd and 3rd by A. La Notte; the 4th and 5th by A. Fraddosio, N. Pecere, and M.D. Piccioni; the 6th is by all the authors.

F. Diosono (✉)
Institut für Klassische Archäologie, Ludwig-Maximilians-Universität München, Munich, Germany

A. Fraddosio · A. La Notte · N. Pecere · M. D. Piccioni
Politecnico di Bari, Dipartimento di Architettura, Costruzione e Design, Bari, Italy

L. Pecchioli (ed.), *Archaeoseismology*, Natural Science in Archaeology,
https://doi.org/10.1007/978-3-031-28303-1_7

7.1 Podium Religious and Political Semantics

The podium is one of the distinctive characteristics of temple architecture of the Etrusco-Italic tradition,[1] appearing in the Etruscan sphere during the first half of the sixth century BC.[2] During the archaic period, the Etruscan-Latium area saw the development of new specific architectural solutions that had to meet requirements that were above all of a religious nature, connected to the concept of *templum*.

Contrary to what modern languages may suggest, the term *templum* in Latin does not mean 'temple' as now the English word and so neither indicate the building in which the deity resided, which was instead the *aedes*. In describing the *templum*, Varro states *omne templum esse debet continuo septum nec plus unum introitum habere*.[3] As regards this study, it must be stated that the *templum* in itself does not correspond with a religious building,[4] but it is finished space, with a single entrance that is firstly delimited verbally *ad auspicando* by the augurs, through the *effatio*. Therefore, the *templum* presents itself as a conceptual space before a physical space, identified and delimited according to the augural divinatory system, which the priests create visually during the ritual and at the same time define *concepta verba*,[5] tracing an orientated space divided by two perpendicular axes into four sectors. According to Linderski, the celestial sphere of the *templum in caelo* (in the air-celestial space dimension) and the terrestrial sphere of the *templum in terra* (in the spatial dimension of the earth's surface) constitute a single entity that is not just conceptual but also material.[6]

This religious concept of the inaugurated, consecrated, and organized space naturally influenced the architectural structure of the Etrusco-Italic temple.[7] According to P. Gros,[8] it can be said that there is a connection between the *templum* as a religious concept and the *podium*, which would represent its architectural manifestation. In fact, the podium delimited the space on which the *aedes* stood, isolating it from the surrounding area and placing it on a higher level, thus enhancing its visibility; furthermore, the podium usually had single access. Therefore, despite the diverse hypotheses formulated on its origins,[9] the podium is the architectural expression of the cult functions associated with the building, which, in reality, represents the materialization of a ritual also described in the Latin sources.[10] Certainly, the subject is far more complex than it is possible to deal here, suffice to say that some Roman religious buildings did not have a *podium*. Indeed, Varro specifies that *templa* should not be confused with consecrated spaces because, for example, the *Curia Hostilia* (the earliest meeting place of the Roman Senate) was a *templum* but not a sacred place; this confusion has been generated by the fact that *in urbe Roma pleraeque aedes sacrae sunt templa*. On the other hand, Aulus Gellius specifies that not every *aedes* was necessarily considered a *templum*, as in the case of the temple of Vesta.[11] Regarding the cult of the Bona Dea, it has been suggested that the absence of podium is due to the fact that these were not inaugurated *templa* on which the *aedes* of public cults stood. Rather they were buildings belonging to the private and domestic sphere of female cults (as indeed were Vesta or Bona Dea), without podium or with a base that is not

---

[1] Gros (2001: 136); Franchetti Pardo (2006: 216).

[2] Bonghi Jovino (2009: 5–8).

[3] Var. *L.* 7.2: each *templum* must have a continuous delimitation and only one access.

[4] On this see also Catalano (1978: 467–477); Castagnoli (1984); Scheid (2009: 67).

[5] Rykwert (1976: 45); Ghey (2005: 109–118). *Concepta verba*: with precise and established words.

[6] Linderski (1986: 2278–2279). See also: Short (2008: 126).

[7] D'Alessio (2010a: 53).

[8] Gros (2001: 134–136). See also Stamper (2005: 10); Coarelli (2011: 51).

[9] Potts (2011: 44–49); Bonghi Jovino (2009: 3–4).

[10] Coarelli (2011: 53).

[11] Gel. 14. 7.7.

inaugurated, defined as *sacraria* or *penetralia* and not *aedes*.[12] Two Roman temples, built during the second half of the second century BC in the southern part of the Campus Martius, are without podium but they have a *crepido*: directly inspired by Hellenistic models both in their plan and the use of marble as a construction material, the temple below the church of S. Salvatore in Campo and the one of *Hercules Olivarius* in the *Forum Boarium* are expressions of the first generation of architects and artists of Greek origin who arrived in Rome following the destruction of Corinth.[13]

Another often underemphasized aspect of ancient architecture is that it was also a form of communication, a visual non-verbal language, which, however, could be more easily understood by its contemporaries than by us. One example is given precisely by the form of the *podia* in the Latin colonies built by Rome in central Italy between the late fourth and third century BC. The type of double-rounded moulding (the so-called double *echinus* or double Etruscan round) that decorated them was completely anachronistic to the architectural style of the period in which they were built, but this effect was probably a carefully considered choice.[14] The aim was to recreate in the newly founded cities of Fregellae, Sora, Aesernia and in the *Forum* of Villa San Silvestro at Cascia an urban landscape familiar to the colonists of Latium. They were to go and live there. Therefore, the temples were built to look like those of the classical period of the Latin cities, such as Ardea and Praeneste, where the podium mouldings, built in the sixth century BC, had counterposed rounded mouldings too. The architecture, in this example, became a tool for the construction of an identity for the new Latin colonies.

Given what has been discussed thus far regarding the religious aspects and polysemantic characteristics that the podium of a Roman temple seems to cover, this paper aims to respond to the question of whether all of that is compatible with not just the architectural but also the structural requirements that such a building could have. Could the podium, together with its foundations, have a role in the protection of the temple that was built on it? The question stems from the fact that the study and architectural analysis of the diverse variants identifiable in the *podia* and foundations of the temples of central Italy has led to the belief that, in some cases, solutions were present that were not linked to providing support for the structures or can be identified as requirements connected with ritual. In particular, we refer to the case of the grid podiums (Fig. 7.1). One of the possible hypotheses we have formulated is that this specific mode of construction had the properties of a 'seismic shield', which protected the sacred building. To verify this theory, a multidisciplinary team of archaeologists, architects, and structural engineers chose a case study on which to test the hypothesis for the first time: Temple B at Pietrabbondante (Isernia), in the modern Italian region of Molise, characterized by high levels of seismic activity.

7.2 Podia of Etrusco-Italic and Roman Temples: Architecture and Construction Techniques

From the architectural point of view, the podium always presents as a platform that isolates and elevates the building above. In literature, the term podium is used to indicate both the base of the Etrusco-Italic temple and, more generally, any raised part of the building provided with steps to overcome the change in height from ground level.[15] More precisely, C. R. Potts has proposed to use the term 'podium' to identify the substructure of a building that occupies all or almost all of the upper surface. At the same time, in the case of a large discrepancy between the dimensions of

[12] Marcattili (2010: 22–23). On the Bona Dea, see Diosono (2019b) with earlier bibliography.

[13] Gros (1976, 2001: 141–143).

[14] Diosono (2016, 2019a).

[15] Ginouvès (1992: 13–14). About the use in the Hellenistic architecture: Von Hesberg (1994). About the Sicilian 'pseudo-podia': Fuduli (2015: 325–326).

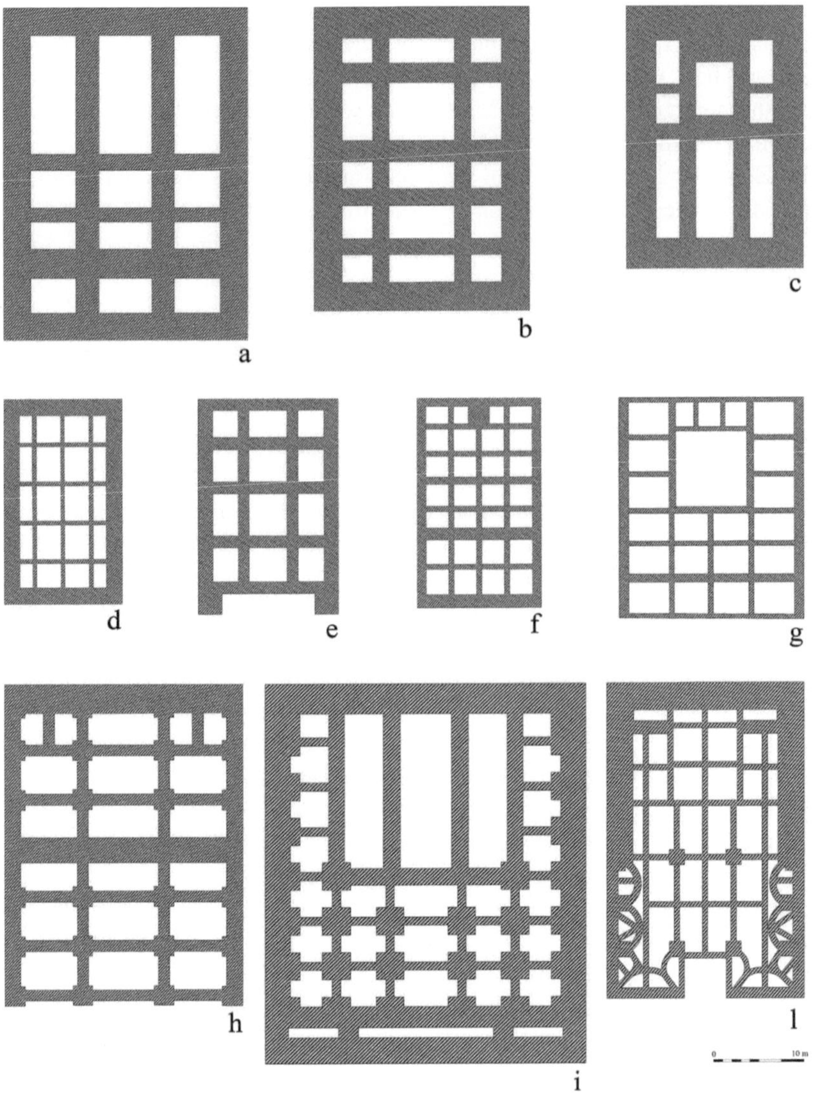

Fig. 7.1 Schematic drawing of grid podium structures. (**a**) Roma, Temple of Castor and Pollux (fifth century BC); (**b**) Pyrgi, Temple A (fifth century BC); (**c**) Pyrgi, Temple B (end of sixth century BC); (**d**) Pompei, Temple of Apollo (second century BC); (**e**) Luni, Capitolium (second century BC); (**f**) Grumentum, Temple D (first century AD); (**g**) Leptis Magna, Temple of Liber Pater (first century AD); (**h**) Terracina, Forum temple (end of first century BC–beginning of first century AD); (**i**) Verona, Capitolium (first century BC); l: Pietrabbondante, Temple B (end of second century–beginning of first century BC) (drawing by A. La Notte)

the two elements, the lower structure takes the name of platform or base.[16]

Whether one regards it as a simple base or an inaugurated structure at the base of a temple, the podium had an important function as a substructure, increased the visibility of the *aedes* (and from the *aedes*), and provided a level horizontal surface, even in the case of uneven terrain.

Despite the constant use in sacred Roman architecture, Vitruvius dedicates little space to the podium in his works, indicating only its constituent parts from the bottom to the top.[17] As the author describes, the podium is characterized by several architectural elements and decorations, which varied over time.[18] Furthermore, podiums differ in plan, materials, and construction technique.

[16] Potts (2011: 41–42); this distinction is also accepted by Bonghi Jovino (2009: 4–5).

[17] Vitr. 3. 4. 5. Potts (2011: 41). About the specific terms: Ginouvès (1992: 13–14); Gros (1997: 326–327), ref. 155.

[18] About the evolution of the base and crown mouldings: Shoe Meritt and Edlund Berry (2000); Edlund Berry (2008); Gros (2001): 147–148.

In most examples, the podium appears as a continuous parallelepiped on whose façade, usually the short side, a flight of steps provides access to the temple.[19] The staircase, varying in size, is frequently incorporated into the body of the podium and flanked by two projecting wings. Otherwise, it is external to the podium and simply abutting the front.[20] Furthermore, in some cases, the steps interrupt the top of the stylobate and are positioned between the front columns due to lack of space and highlight the single columns, which thus appeared raised on pedestals.[21]

From the late Republican age onwards, there is a growing tendency to evolve the morphology of the podium and the stairways, to accommodate spaces intended for specific political and judicial activities.[22] Furthermore, the use of 'hollow substructures' within the podium is important. In this case, the podium has a nucleus divided into a modular succession of chambers, sunken-floored or totally above ground level. These chambers could be either inaccessible,[23] or accessible from the exterior through doors and lit by windows or slits.[24] The functional use of these spaces was usually connected with religious activities (*favissae*),[25] but in some cases, commercial activities or deposits of public goods are attested too.[26] Favoured by the use of concrete and vaulted coverings, this architectural solution widespread from the second half of the second century BC onwards, not only in temple podiums, but also in the complex hollow substructures of the sanctuary spaces in Lazio and Campania.[27]

Regardless of its complexity, the structure of the podium contributes to transmitting the entire load of the upper building to the ground, thus ensuring elevated and enduring stability. Contrary to the perimeter, which is architecturally defined, the internal walls are generally hidden and often do not foresee a strict distinction in the construction technique used in the parts below floor level (the foundations proper) and the upper parts (substructures).[28] The structure of the podium can be classified based on construction techniques, materials used, their depth and therefore, the load-bearing capacity of the terrain, and lastly, the plan.[29]

Regarding the plan, the following forms are recurring:

- Continuous structures (or strip structures),[30] with linear and generally homogeneous walls nearly repeating the plan of the upper building.[31]
- Continuous structures are combined with isolated piers when, in addition to the strip foundations under the central *cella* and under the perimeter walls and colonnades, isolated

---

[19] Exceptions are the temples with a *cella traversa* (Marcattili 2017), as well as the use of lateral steps such as the temple of the Ara Rotonda at Ostia (Pensabene 2007: 53–63), or steps at the rear of the podium, like the temple of *Sol Indiges* at *Lavinium* (Jaia 2017).

[20] Pensabene (1991): Figs. 4–6. On the measurements and proportions of the steps: Vitr. 3. 4. 5.

[21] E.g. the Metellan Temple of Castor and Pollux in Rome (Nielsen 1992: 112), the Temple of *Divus Iulius* in Rome (Montagna Pasquinucci 1973), the Temple of Augustus and Rome in the forum of Leptis Magna (Livadiotti and Rocco 2005: 193).

[22] Lippolis (2016: 140–141). E.g. the Temple of *Venus Genetrix* (Delfino 2014) and the Temple of *Divus Iulius* (Montagna Pasquinucci 1973) in Rome, the *Capitolium* in Pompeii (Lippolis 2016: 125–127), or the Temple of Augustus and Rome at Leptis Magna (Livadiotti and Rocco 2005: 197–203).

[23] E.g. temple A at *Minturnae* (Bankel 2016: 81–82). Valenti 2016: 51, type C1.

[24] E.g. the *Capitolium* of Pompeii (Lippolis 2016: 122–123, 139–140), the Temple of Castor and Pollux (Nielsen 1992); Valenti (2016: 51–52), type C2.

[25] Valenti (2016: 52–53).

[26] Potts (2011: 47–48).

[27] D'Alessio (2010b, 2011). For the use of vaulted underground spaces in the Greek area: Lauter (1986: 60–63); Gros (1997: 323), ref.146.

[28] Often, it is impossible to distinguish the *euthynteria* from the internal walls of the podium.

[29] Giuliani (2006: 161–180); Valenti (2016: 51–52).

[30] Ginouvès (1992: 10).

[31] E.g., the mid-Republican temples of the Sanctuary of Diana at Nemi (Diosono et al. 2019) or the Temple of Juno at *Gabii* (Almagro Gorbea 1982).

piers are positioned to support the inner columns of the pronaos.[32]

- Platform structure,[33] when the podium appears as a massive nucleus without internal cavities, often built in *opus caementicium* and reinforced with masonry piers below the columns and in the points of significant stress, where necessary.[34]
- Grid structure (or with cross-linked foundation piers),[35] when the continuous substructures and eventually the reinforcing piers are connected by transverse and longitudinal walls, forming an orthogonal grid. These connecting walls, however, do not correspond with any vertical building structures.[36]

Podiums with an internal grid structure, in particular, were often used in Etrusco-Italic temples from the archaic period onwards, as attested by the cases of the *Capitolium*[37] and the Temple of Castor and Pollux[38] in Rome (Fig. 7.1a) and the temples A and B at Pyrgi[39] (Fig. 7.1b, c). As these examples show, the *opus quadratum* grid presents hierarchically distinct walls: those supporting a more significant load, along the perimeter of the podium and underneath the walls and columns of the temple are wider, while the walls exposed to less stress, that is the connecting cross walls forming the grid, were narrower. More homogeneous walls, both in *opus quadratum* and *opus caementicium*, characterize several late Republican buildings, such as the temples of Apollo at Pompeii[40] (Fig. 7.1d) and the *Capitolium* of Luni (Fig. 7.1e)[41]; moreover, in the imperial period this technique was used for the podium of the Temple of *Liber Pater* at Leptis Magna (Fig. 7.1g)[42] and Temple D at *Grumentum* (Fig. 7.1f).[43]

A further plan of the grid structure is the one in the podium of the forum temple at Terracina (Fig. 7.1h),[44] the *Capitolium* at Verona (Fig. 7.1i),[45] and the temple B of Pietrabbondante (Fig. 7.1l). In these examples, the grid is not homogeneous but with thick walls around the perimeter and below the temple's *cella*, large quadrangular piers under the columns, and narrow walls linking these elements. Constructed in this form, the podium seems to respect Vitruvius's prescriptions for building solid foundations for a temple.[46] According to the author:

> The foundations of these works should be dug out of the solid ground, if it can be found, and carried down into the solid ground (. . .) Above ground, let walls be laid under the columns, thicker by one-half than the columns are to be, so that the lower may be stronger than the higher (. . .). And the projections of the bases should not extend beyond this solid foundation. The wall-thickness is similarly to be preserved above ground likewise, and the intervals between these walls should be vaulted over (*concameranda*), or filled with earth rammed down hard (*festucationibus*), to keep the walls well apart.[47]

The Vitruvian text seems to refer to the practice of sub-dividing the nucleus of the podium into underground spaces covered by a masonry vault (*camerae*) which, as mentioned above, could either be inaccessible and function only as substructures, or accessible and destined for a variety of uses, not necessarily linked to the sacred sphere. Alternatively, the practice of filling the empty spaces between foundation structures

32 Ginouvès (1992: 11). E.g. the Temples of *Juno Sospita* at *Lanuvium* (Bendinelli 1921), the Temple of *Sol Indiges* at *Lanuvium* (Jaia 2017), the Republican Temple of *Apollo Sosianus* in Rome (Viscogliosi 1996). Valenti 2016: 51 type A1, A2.

33 Ginouvès (1992: 10–11).

34 E.g. the Temple of Hercules at Cori (Brandizzi Vitucci 1968:78); Temple of *Apollo Palatinus* in Rome (Zink 2012); Temple of Apollo in *Peltuinum* (Bianchi 2012); Valenti 2016: 51 type B.

35 Ginouvès (1992: 11), note 30.

36 Bianchini (2010: 179–182).

37 Cifani (2008: 80–109).

38 Nielsen and Poulsen (1992: 61–79).

39 Colonna (1970).

40 Rescigno (2016: 42, fig. 7).

41 Forte (1991: 106–110).

42 Masturzo (2005: 112–113).

43 Fusco et al. (2009: 188–191).

44 Valenti (2016).

45 Cavalieri Manasse (2008). About the podium: 76–83.

46 Adam (1988: 115–117).

47 Vitr. 3.4.1 (transl. Gros 1997); Gros (1997: 251–253).

with soil (*festucationes*) was widespread among the Roman-Italic temples.[48] As attested in the temples of Pyrgi and Rome,[49] the spaces in the foundation were filled gradually and in several stages between the construction of each course of the walls. In this way, the internal fill also functioned as a useful levelled floor for the positioning of the blocks. However, the soil volume exerted an orthogonal pressure on the podium's perimeter walls. In order to decrease its intensity, soil with a clay matrix was often used in the fill, mixed with discarded materials from the building site and any rubble from the demolition of earlier buildings.[50] In fact, the clay tended to compact and solidify thanks to the absorption of humidity from below ground, thus decreasing the pressure on the perimeter walls and, at the same time providing greater resistance to the vertical load of the paving slabs above.[51] The grid of walls inside the podium also improved the structure's resistance as the fill was subdivided into a series of small compartments, thus decreasing the pressure on the exterior walls.

7.3 Case Study: The Podium of Temple B at Pietrabbondante

If it is clear that podiums with an internal grid structure contribute to the stability of the building, however, Vitruvius does not explain in detail the reasons. It is possible that benefit was meant not only for the stability of the substructure, with respect to the pressure from the internal fill, but also in the case of external stresses such as earth tremors. This research aims to investigate whether this specific podium structure could contribute to the entire temple building's resistance, especially against earthquakes. To undertake an investigation of this nature, Temple B at Pietrabbondante (Isernia) was chosen for an in-depth study.

The building is located on the slopes of Monte Saraceno in the central Apennines, a region with a complex geological structure and rich in materials suitable for construction, as the temple attests. The area is mainly characterized by alternating different types of marl, red clays, and fine limestone breccias, highly recrystallized.[52] Furthermore, massive beds of sandstones prevailingly composed of subrounded quartz granules, layers of clay of variegated colour and whitish recrystallized limestones (calcirudites, calcarenites, and calcilutites) are diffused too.[53]

The area is characterized by a mid-high level of seismic risk, where high-intensity earthquakes can be frequent. It is close to the Piana di Boiano, one of the central-southern Apennines's most effective seismogenic systems and characterized by various fault systems.[54]

Although the historical sources are relatively poor of information regarding the Republican period, at least three critical earthquakes are attested in the area around Pietrabbondante between the first and fourth centuries AD.[55] In this period, tremors of high intensity and destructive effects on the buildings affected the *Samnium* area and specifically the city of Isernia.[56] A more significant number of medium-high intensity earthquakes are instead documented with continuity from the Middle Ages to nowadays, suggesting a similar frequency even in the previous period, despite the scarce documentation.[57] The high seismic risk, therefore, identifies the area as particularly interesting for the investigation.

[48] In the Ionian-Asia Minor area, a grid structure is used in the internal nucleus of the Great Altar at Pergamon (Schrammen 1906: 12–14, Tav. II).

[49] Colonna (1970: 38); Cifani (2008: 87, 120).

[50] Architectural fragments are often found in the fill of temple podiums that belonged to the preceding phase of the sacred building, deliberately demolished in order to modernize its forms or following traumatic events such as fires or earthquakes.

[51] Cifani (2008: 120); Bianchini (2010: 181), ref. 206.

[52] Sgrosso and Naso (2011: 46).

[53] Sgrosso and Naso (2011: 49–52).

[54] The earthquakes intensity can exceed X degrees on the scale Mercalli–Cancani–Sieberg (MCS). Sgrosso and Naso (2011): 101–102.

[55] Guidoboni et al. (1994, cat. 104, 144, 159).

[56] D'Apollonio and Damiani (1991).

[57] Guidoboni et al. (2018).

The Temple B of Pietrabbondante is included within the main sanctuary of the *Samnites Pentri*,[58] but it stands out among the numerous buildings of the sanctuary,[59] as the fulcrum of a monumental complex that also includes a theatre in front of it and two colonnaded porticoes on both sides of the *aedes*.[60] The complex was built between the second half of the second century and the first decade of the first century BC, following the Hellenistic models and, in particular, the typology of the 'Roman Theatre-Temple'.[61] The presence of buildings around the temple could have helped protect it from the impact of seismic shocks. However, in this study, they are not taken into consideration, as Temple B is analysed as a case study for its particular foundation structures, deferring an in-depth analysis of the architectural context for further research. The entire temple building rests on a large rectangular podium (35 m long, 22 m wide, 3.55 m high), decorated by moulded base and crown,[62] accessible on the short side via a long staircase flanked by two projecting wings.

The distinctive feature of the podium is the complex network of walls in its interior, differing in form, construction technique, and structural function.[63] The substructures of the walls and columns of the temple are built in *opus quadratum,* in large blocks of fossiliferous limestone (Fig. 7.2a). More precisely, below the *cellae* and the podium perimeter, there are long continuous walls, whereas below the columns there are isolated piers, all reaching to a depth of c. 1.3 m below the paved floor at the sides of the temple. The second network of walls in the nucleus of the podium connects the previous ones but has no relationship with the upper structures (Fig. 7.2b). These walls are narrower (c. 0.60 m) and are not set as deeply into the ground as the actual foundations, which they only touch, but without any solid connection. They are built in *opus caementicium*, with irregular stones and reused materials from earlier buildings bonded with lime mortar. These walls are set to form an almost regular orthogonal grid and divide the podium into several rectangular cells filled with rubble and clayey soil, resembling the *festucationes* described by Vitruvius. In the proximity of the columns and the podium walls in the *pronaos* area, these concrete walls assume more complex forms, almost semi-circular or parabolical, reinforced by one or two walls converging at the centre of the curve. The particular curved form can be interpreted as an element providing a better counterthrust to the pressure from the internal fill.[64] Semi-circular walls, in fact, are frequently used in terracing structures and substructures,[65] although rarely attested in temple architecture.

Therefore, the podium of Pietrabbondante appears as a complex construction responding to structural requirements thanks to the form, arrangement, and technical characteristics of its walls. The period and the cultural context of the temple suggest that the architecture of the podium may have benefited from technical know-how that had been enriched through time via positive construction experiences. In particular, the internal grid structure could have been chosen because of the numerous advantages offered to the building's stability, including a possible contribution to its anti-seismic resistance. The empirical verification of a general increase in the stability of the structures, an improvement in their static and dynamic equilibrium, and thus in their durability through time, even in the event of an earthquake,

[58] La Regina (1989). On the history of the research on the site: La Regina (2014: 161–169).

[59] Strazzulla (1973); La Regina (1978).

[60] Di Marco (1996); La Regina (2014: 173–188).

[61] Hanson (1959). On the use of this combination in middle-Italic sanctuaries see the following (with preceding bibliography): Gros (2009); D'Alessio (2010b, 2011); Coarelli (2012: 187).

[62] The base presents a cavetto followed by a *cyma reversa*. The crown, instead, is formed by an half-round astragal followed by a quarter-round ovolo and a cavetto, forming together a *cyma recta* (Gros 2001: 140).

[63] Strazzulla (1973: 25).

[64] Bianchini (2010: 181–182).

[65] E.g. the niches and apses in the sanctuary of Diana at Nemi (Ghini 2013), the so-called Villa of Pompey at Albano Laziale (Jaia 2019) and the eastern part of the so-called Villa of Nero at Anzio (Di Mario and Jaia 2009: 72–73, figs. 27, 28).

a

b

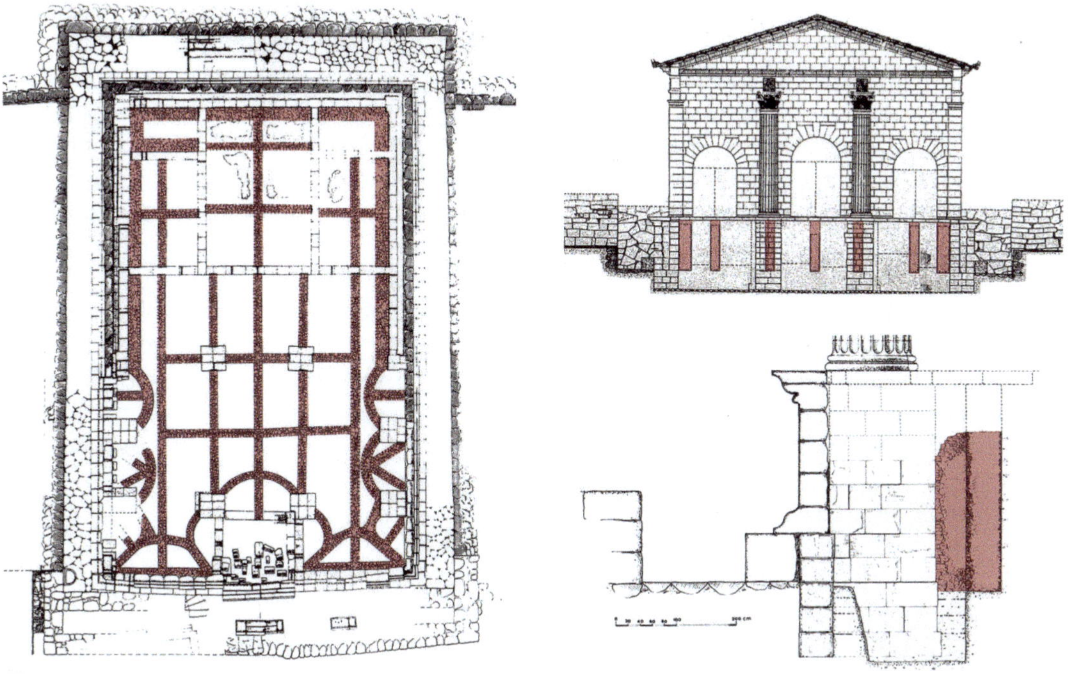

Fig. 7.2 Pietrabbondante, Temple B, planimetric survey of the remains, hypothesis of reconstruction of the elevation and detail of the podium. In red, the podium structures in *opus quadratum* (**a**) and in *opus caementicium* (**b**) (drawing by A. La Notte, after La Regina 1989)

could have determined the success of this construction technique.

7.4 Seismic Metamaterials and Seismic Archaeological Meta Structures

To understand how the particular shape of the foundations of Temple B of Pietrabbondante can offer advantages in terms of seismic protection, it is necessary to introduce the concept of seismic metamaterials (SMs) and, in particular "seismic archaeological metastructures".

From a historical point of view, after some attempts between the end of the 19th century and the 1960s, one of the main pioneering theoretical studies on materials known today as metamaterials dates back to 1968 and was due to the Russian physicist V. Veselago.[66] In his seminal paper, Veselago showed that an ideal material characterized by both negative values of the electrical permittivity and the magnetic permeability would have a negative refractive index for electromagnetic waves, contrary to natural materials. This paved the way for the development of artificial materials, called "metamaterials", characterized by properties not achievable with natural materials of blocking, absorbing, enhancing, or bending electromagnetic waves. These properties are due to their design, and in particular to their shape, geometry, size, orientation, and arrangement.

Only in the late 1990s did it become possible to design and properly manufacture metamaterials due to the pioneering work of J. Pendry.[67] Later, the properties of metamaterials were exploited for creating an invisibility cloak, i.e., a special device capable of making invisible an object, first realized by D. Schurig et al.[68] Technical details, based on the concepts of gauge conditions and coordinate transformations can be found in the contribution of W.S. Gan.[69]

In parallel with the research on electromagnetic metamaterials, in the 1990s, the research has been extended to the acoustic and elastic fields, and periodic structures capable of manipulating acoustic waves, called "phononic crystals," were studied. This was also motivated by the similarity of the structure of some equations ruling the propagation of electromagnetic waves and acoustic or mechanical waves (see the considerations about the gauge invariance in W.S. Gan's work[70]). Therefore, the concept of cloaking was extended to the acoustic field by Milton et al. in 2006,[71] by Leonhardt in 2006,[72] and Cummer and Schurig in 2007[73] using the coordinate transformations approach.

In particular, negative refraction is a special case of coordinate transformations resulting from the negative mass density and the negative bulk modulus of an acoustical metamaterial. The latter can also produce phononic crystal's band gap, i.e. frequency bands where transmission is forbidden.

In the last decade, numerous studies have been carried out on elastic metamaterials capable of controlling the propagation of elastic waves in a solid medium.

In this context, a fundamental contribution by S. Brûlé et al.[74] paved the way for the application of the concept of metamaterial in a different engineering field, i.e., seismic protection of structures, giving birth to the so-called "seismic metamaterials" (SMs).

Also, the concept of seismic cloaking using metamaterials has been explored by several authors[75] to study whether it is possible to surround a structure with a suitable metamaterial capable of totally shielding the structure from

[66] Veselago (1968: 509–514).

[67] Pendry et al. (1999: 2075–2984).

[68] Schurig et al. (2006).

[69] Gan (2018, 2019).

[70] Gan (2007).

[71] Milton et al. (2006).

[72] Leonhardt (2006).

[73] Cummer and Schurig (2007).

[74] Brûlé et al. (2014).

[75] Chiroiu et al. (2014); Finocchio et al. (2014); Khlopotin et al. (2015); Colombi et al. (2016); Colquitt et al. (2017); La Salandra et al. (2017); Maurel et al. (2018); Muhammad et al. (2019); Ungureanu et al. (2019); Colombi et al. (2020); Mu et al. (2020); Muhammad et al. (2020).

earthquakes or at least substantially reducing seismic effects.

A further recent step of this research line concerns the possibility that existing cellular structures, like the foundations of some special buildings, could act as a seismic shield. In particular, S. Brûlé et al.[76] and Brûlé and Guenneau 2021[77] observed that certain ancient structures buried in the ground, or placed on top of it, can determine a favourable behaviour against seismic waves. This observation emerges from the clear analogy between the shape of microscopic objects acting as a cloak for electromagnetic waves, and the shape of the foundations of a Roman amphitheatre, with the same regular patterns of cells, obviously at very different scales.

The idea could help in explaining the resilience to seismic events over centuries shown by these large and complex ancient buildings, often placed in very active seismic zones, and gave rise to the study of seismic archaeological metastructures and metamaterials. Of course, the goal of designing seismic shielding foundations has not been intentionally achieved by Romans, although the observation of the seismic effectiveness could have driven the development and the selection of particular shapes for the amphitheatres and their foundations over centuries. These studies open new, surprising horizons for understanding ancient constructions.

7.5 Seismic Shielding Effect of the Pietrabbondante Temple B Foundations

The above-recalled suggestion given by Brûlé et al.[78] about the possibility that the foundations of some ancient large structures could act as seismic shields motivates the analyses performed in the present research.

Even if the morphology of the foundations of a rectangular-plan temple cannot have a very strict similarity with the morphology of a cloaking device for the light that has been highlighted for theatres and amphitheatres foundations, the very particular cellular shape of the foundations of Pietrabbondante Temple B is worth to be investigated in terms of the capacity of mitigating seismic effects on the superstructure.

Indeed, Pietrabbondante Temple B foundations (Fig. 7.3, left) are shaped like a cell net, realized also by employing curved foundation walls. This shape apparently does not meet structural functions, as there are no walls or columns of the superstructure on many cell walls. Moreover, the cells do not have soil support functions, because the perimeter of the foundations is disconnected from the surrounding soil.

Since Pietrabbondante is a medium-high risk area from aseismic point of view, a possible hypothesis is that the special shape of the foundations is the result of the evolution of local constructive techniques, leading to select those shapes that over time shown to be more effective in protecting the superstructure to earthquakes. If this hypothesis is true, the foundation shape under investigation can be viewed as an example of archaeological seismic metamaterial, whose design has been most probably determined by empirical observations of structural effects after earthquakes.

To study the seismic shield capacity of the foundations of Pietrabbondante Temple B, numerical simulations have been performed using an advanced Finite Element (FE) code, Comsol Multiphysics.[79] This code can effectively simulate wave propagation phenomena in different kinds of media, and it is usually employed for numerical analyses in metamaterials.

In particular, the effectiveness of the special cellular shape of Pietrabbondante Temple B foundations is studied by comparing seismic effects on the top of the foundations between the case of the actual foundation's shape and the case of an ideal 'ordinary' foundation system composed of walls placed only below the structures of the temple.[80]

---

[76] Brûlé et al. (2019).

[77] Brûlé and Guenneau (2021).

[78] Brûlé et al. (2019).

[79] COMSOL Multiphysics.

[80] On this topic, see also Diosono et al. (2022).

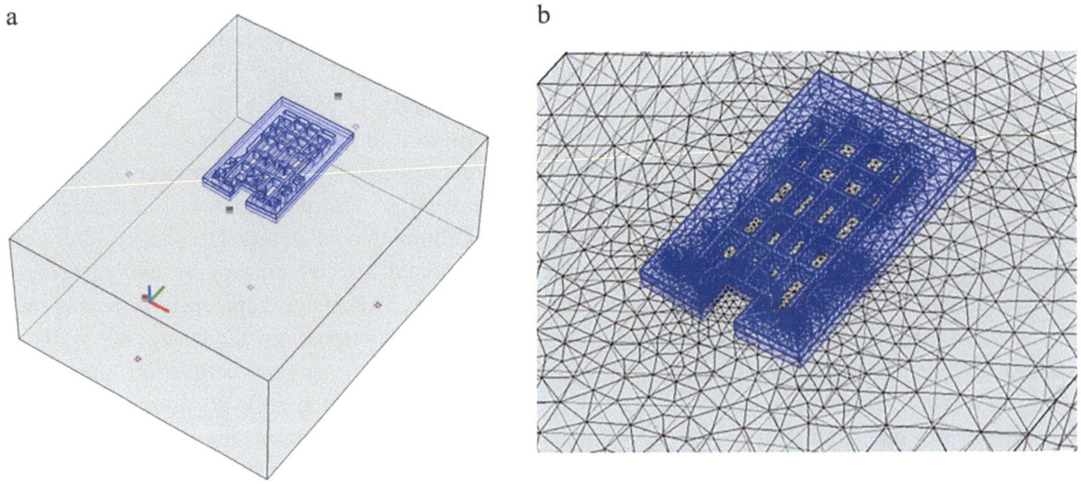

Fig. 7.3 (**a**)FE model of the podium structures and the surrounding soil, elaborated on the basis of the surveys (Fig. 7.2); (**b**) detail of the 3d mesh (drawing by N. Pecere)

From a geometrical point of view, the analysed FE model is composed of a parallelepiped measuring 100 m x 80 m in plan and 40 m in the vertical direction, on which the temple foundations are placed (Fig. 7.3, top right). For structural simulations, a simplified geometry of the foundations has been considered. The parallelepiped represents the foundation soil; its dimensions have been chosen as a reasonable compromise between the need for defining a sufficiently large portion of the soil and for reducing as much as possible the dimension, and consequently the computational costs. Indeed, these simulations are very burdensome in terms of computational time and resources. Notice that for matching the actual on-site conditions, the lower part of the foundation walls is immersed in the soil for 1.5 m, whereas the upper part, a height of 1.5 m, is above the ground.

To have representative results, the mesh of the solid domain representing the soil in the FE analysis has been carefully designed, coarser far from the temple foundations, and progressively more refined near the foundations (Fig. 7.3, downright). The latter has also been discretized with a quite refined mesh. In particular, second-order solid finite elements (20-node hexahedral, etc.) have been employed for the discretization of the soil and the foundations. In spite of the aforementioned simplifications, the model is complex since it is composed of 77,832 solid finite elements.

From the mechanical point of view, for the present analysis, the clay soil has been represented as a linear elastic and isotropic material having the following mechanical properties: Young's modulus $E = 50$ MPa and Poisson's coefficient $n = 0.28$. Further, foundation walls have been modelled as linear elastic and isotropic, with Young's modulus $E = 2000$ MPa and Poisson's coefficient $n = 0.20$. In both cases (soil and masonry of the foundation), experimental information is unavailable; therefore, the assumed mechanical parameters represent reasonable values for the considered materials.

A relevant issue is the representation of seismic actions in the numerical model. Indeed, it is well known that an earthquake involves different kinds of waves: body waves, like P and S waves, and surface waves, like Rayleigh and Lamb waves. Also, it is known that the most intense effects on constructions are generally due to Rayleigh surface waves. Therefore, to simplify the calculations, only the latter kind of waves has been simulated by introducing a point source on the surface of the soil domain, placed along the longitudinal axis of the temple foundations, at a

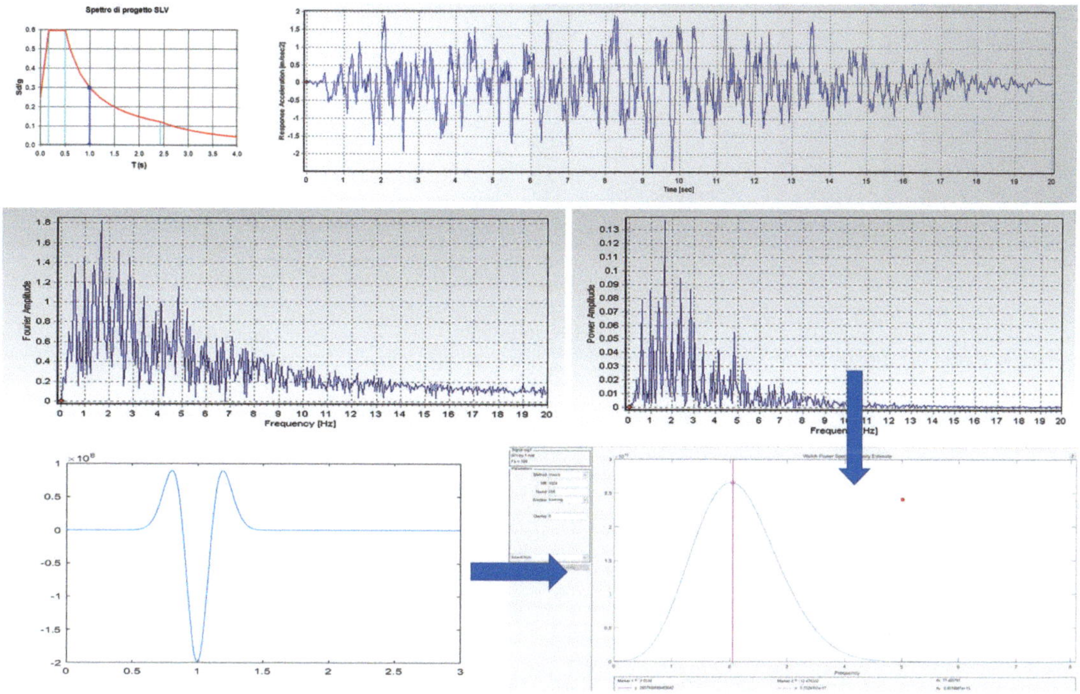

Fig. 7.4 Ricker wavelet for simulating the seismic source (graphic by N. Pecere)

distance of 50 m from the front edge of the foundations.

To simulate with the FEM model an ideally infinite medium, appropriate boundary conditions of low reflectivity have been applied on the lateral surfaces and on the lower surface of the spatial domain. Moreover, for representing the seismic signal employed as the point source, according to the more advanced literature on seismic simulations, a Ricker wavelet[81] is employed. Indeed, Ricker wavelet allows a broad wave spectrum that can better match that of real earthquakes.

In particular, to determine the features of the Ricker wavelet to be introduced in the model, the spectrum representing seismic actions for structural design in Pietrabbondante according to the Italian building code[82] has been determined. Starting from the spectrum, a spectrum-compatible acceleration time history has been obtained employing the software SIMQKE_GR, specifically aimed at this purpose. Then, the power spectral density of this signal was evaluated, and finally, the Ricker wavelet parameters were determined such that the Ricker wavelet had the main features of the power spectral density as close in similarity to that of the actual seismic acceleration signal (Fig. 7.4). The duration of the employed Ricker wavelet is 1.00 s, and the duration of the wave propagation simulations is 3.00 s.

Three different models have been analysed: – in the first model, only the soil, to study the soil motion in positions corresponding to particular points of the temple foundations plan; – in the second model, the temple foundations; finally, in the third model, a simplified foundation system (called ‘minimal’ foundation system), with walls present only below the superstructure of the temple. The comparison of the results obtained for the latter two models allows us to understand the role of the special cellular morphology of the temple foundations in reducing the intensity of earthquake effects on the supporting structure.

[81] Yanghua (2015).

[82] Italian Ministerial Decree 17.01.2018 *‘Aggiornamento delle Norme Tecniche per le Costruzioni’ (NTC)*.

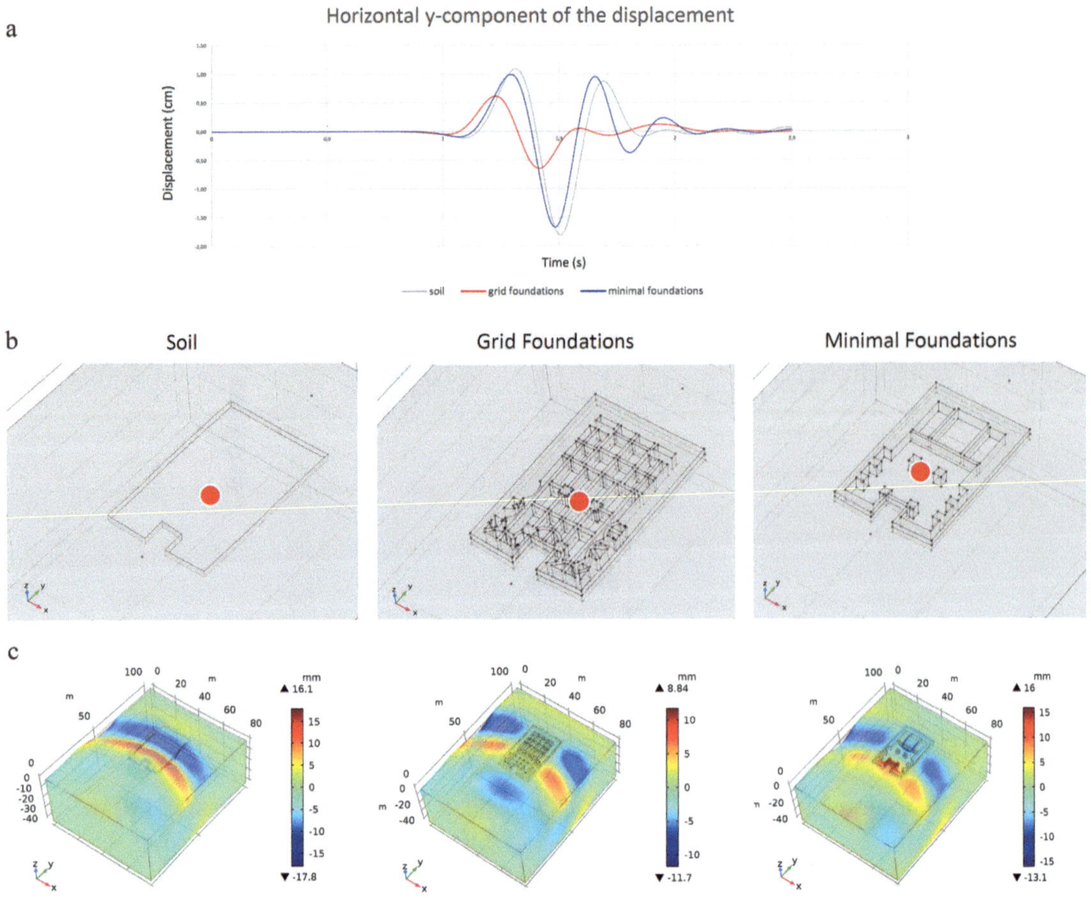

Fig. 7.5 (**a**) Timeline history of the horizontal displacements in y direction of a fixed point in case of soil without the foundations (grey line), soil with the temple foundations in their actual cellular shape (red line), and soil with the 'minimal' foundation system (blue line); (**b**) In red, the position of the displacement measurement point in the three analysed cases; (**c**): Simulation of the seismic response in the three analysed cases (drawing by N. Pecere)

It is worth noting that for the study here reported, representing the first attempt to understand if the foundation's shape can make them a seismic metamaterial, the structure of the temple above the foundations is not represented in the model. This simplification does not weaken the relevance of the analysis, focused on the reduction of the seismic signals transmitted from the soil to the superstructure through the foundations.

The performed simulations show that the presence of the temple foundations altered the propagation of seismic waves. In particular, it is observed that the special foundation's grid of Pietrabbondante Temple B leads practically to halve the amplitude of horizontal displacements. The result indicates that the cell foundation system acts as a filter for seismic waves, that is, as a seismic metamaterial, substantially reducing the motion transmitted to the structure of the temple. Moreover, this reduction is not obtainable with a standard foundation system, like the minimal foundation system. Indeed, in the latter case, the amplitude of the displacements above the foundations is the same order as that of the soil.

Figure 7.5 illustrates the above-described findings for a significant point of the foundations placed at the base of a temple's column. In the figure, the diagram on the top displays the time history of the horizontal component of the displacement in the direction of the temple's axis (y-axis). The grey line represents the response of the considered point for the model representing only the soil. In this case, the oscillations of the supposed point reach a maximum amplitude of about 1.44 cm in the positive direction and of about −2.02 cm in the negative order. The red line represents the results for the model, including the temple foundations with their actual cellular shape: the reduction concerning the previous case is marked since now the maximum amplitude is about 0.61 cm in the positive direction and about −0.65 cm in the negative direction. Hence, the special foundation system allows for strongly mitigating seismic effects on the superstructure, lowering more than 58%. Finally, the blue line refers to the model's results with the minimal foundation system. Here, the maximum amplitude of the displacements of the considered point is about 0.98 cm in the positive direction and about −1.64 cm in the negative direction. Therefore, no significant effects in terms of mitigation of seismic effects for the overlying temple are allowed for this foundation system.

7.6 Conclusions

The preliminary analysis here performed confirms that the cellular shapes of the *podium* of Temple B of Pietrabbondante, belonging to the grid foundations group, could play a significant role in reducing seismic effects on the construction.

Since, in ancient times, no seismic knowledge was analytically developed, it could be assumed that this foundation shape has been achieved empirically over time, with the aid of the observation of eventual seismic effects on the superstructure too. In fact, no ancient text is preserved referring to this construction technique with a specific 'seismic shield' purpose; we have observed how Vitruvius, in speaking of techniques for making the foundations of temples more solid, does not explicitly refer to the threat that earthquakes can pose to structures. However, according to the available sources, we do not have precise references for several other construction techniques. It must also be said that for the ancient world, the practical knowledge in construction, which spread with the mobility of the workers, had to have a decisive effect in the context of construction sites of which we are not currently able to evaluate the extent.

The obtained results encourage a further deepening of the study, considering aspects such as the interaction between the foundations and the above temple structure, the influence of the direction of the earthquake and the effects of more complex but realistic seismic excitation, the role of all the wave modes propagating in the soil, including P and S waves. Furthermore, the eventual influence of other constructions surrounding the temple could be considered. Also, a more advanced constitutive modelling of the building materials and masonry techniques can be introduced.

Besides the above-listed improvements, the results of this preliminary study indicate the fruitfulness of the multidisciplinary approach employed, based on the close interaction between different points of view, such as that of the archaeologist, the architect, and the structural engineer.

References

Adam JP (1988) L'arte di costruire presso i Romani. Longanesi, Milano

Almagro Gorbea M (ed) (1982) El santuario de Juno en Gabii, EEHAR, Bibliotheca italica, vol 17. Consejo Superior de Investigaciones Cientificas, Roma

Bankel H (2016) I due templi nel cosiddetto Foro repubblicano di Minturnae. Un rapporto preliminare. In: Valenti M (ed) L'architettura del sacro in età romana. Paesaggi, modelli, forme e comunicazione, Gangemi Editore, Roma, pp 79–88

Bendinelli G (1921) Monumenta lanuvina. Accademia dei Lincei, Monumenti antichi, vol 27. Accademia dei Lincei, Roma, pp 293–370

Bianchi F (2012) Il tempio del Foro. Atti della Pontificia accademia romana di archeologia. Rendiconti 84:287–330

Bianchini M (2010) Le tecniche edilizie nel mondo antico. Dedalo, Roma

Bonghi Jovino M (2009) Alle origini del processo di strutturazione del tempio etrusco. La presenza del podio. Studi Etruschi 75:3–8

Brandizzi Vitucci P (1968) Cora. Forma Italiae, Regio I, vol 5. De Luca, Roma
Brûlé S, Guenneau S (2021) Past, present and future seismic metamaterials: experiments on soil dynamics, cloaking, large scale analog computer and space-time modulations. Comptes Rendus Physique 21(7–8): 767–785
Brûlé S, Javelaud EH, Enoch S, Guenneau S (2014) Experiments on seismic metamaterials: Molding surface waves. Phys Rev Lett 112:133901
Brûlé S, Enoch S, Guenneau S (2019) Role of nanophotonics in the birth of seismic megastructures. Nanophotonics 8(10):1591–1605
Castagnoli F (1984) Il tempio romano: questioni di terminologia e di tipologia. Pap Br Sch Rome 52:3–20
Catalano P (1978) Aspetti spaziali del sistema giuridico-religioso romano. Mundus, templum, urbs, ager, Latium, Italia. Aufstieg und Niedergang der Römischen Welt (ANRW) II.16, vol 1, pp 440–553
Cavalieri Manasse G (2008) Gli scavi del complesso capitolino. In: Cavalieri Manasse G (ed) L'area del Capitolium di Verona. Ricerche storiche e archeologiche. Soprintendenza per i Beni Archeologici del Veneto, Verona, pp 73–152
Chiroiu V, Munteanu L, Ioan R, Moşneguţu V (2014) On the seismic cloaking. Romanian J Acoust Vib 11(1): 31–34
Cifani G (2008) Architettura romana arcaica: edilizia e società tra Monarchia e Repubblica. Bibliotheca archaeologica, vol 40. L'Erma di Bretschneider, Roma
Coarelli F (2011) Le origini di Roma. La cultura artistica dalle origini al III secolo a.C. Storia dell'arte romana, vol 1. Jaca Book, Milano
Coarelli F (2012) L'architettura del Lazio in età tardorepubblicana. In: von Hesberg H, Zanker P (eds) Storia dell'architettura italiana. Architettura romana. Le città in Italia, Electa, Milano, pp 176–207
Colombi A, Colquitt D, Roux P, Guennea S, Craster RV (2016) A seismic metamaterial: the resonant metal wedge. Sci Rep 6:27717
Colombi A, Zaccherini G, Aguzzi A, Chatzi E (2020) Mitigation of seismic waves: metabarriers and metafoundations bench tested. J Sound Vib 485: 115537
Colonna G (1970) Pyrgi: scavi del santuario etrusco (1959-1967). Notizie degli scavi di antichità. Supplemento 24, 1-2Accademia Nazionale dei Lincei, Roma
Colquitt DJ, Colombi A, Craster RV, Roux P, Guenneau S (2017) Seismic metasurfaces: sub-wavelength resonators and Rayleigh wave interaction. J Mech Phys Solids 99:379–393
Cummer SA, Schurig D (2007) One path to acoustic cloaking. New J Phys 9:45
D'Alessio A (2010a) Fascino greco e attualità romana: la conquista di una nuova architettura. In: La Rocca E, Parisi Presicce C (eds) I giorni di Roma. L'età della conquista. Catalogo della mostra, Roma, Musei Capitolini, 13 marzo–26 settembre, vol 2010. Skira, Milano, pp 49–64
D'Alessio A (2010b) Santuari terrazzati e sostruiti italici di età tardo-repubblicana: spazi, funzioni, paesaggi. In: Meeting between Culture, XVII AIAC Congress, Roma 22–26 settembre 2008. Bollettino di Archeologia online 2010, edizione speciale, pp 17–33
D'Alessio A (2011) Spazio, funzioni e paesaggio nei santuari a terrazze italici di età tardo-repubblicana. Note per un approccio sistemico al linguaggio di una grande architettura. In: La Rocca E, D'Alessio A (eds) Tradizione e innovazione: l'elaborazione del linguaggio ellenistico nell'architettura romana e italica di età tardo-repubblicana. Studi miscellanei, vol 35. L'Erma di Bretschneider, Roma, pp 51–86
D'Apollonio V, Damiani P (1991) Cronotassi dei terremoti in Isernia e nel resto del Molise, 346–1986. Marotta, Napoli
Delfino A (2014) Forum Iulium. L'area del Foro di Cesare alla luce delle campagne di scavo 2005–2008: le fasi arcaica, repubblicana e cesariano-augustea. British Archaeological Reports International, Series 2607. Archaeopress, Oxford
Di Marco B (1996) Studio di ricostruzione grafica del complesso cultuale di Pietrabbondante. Bollettino dell'Istituto regionale per gli studi storici del Molise V. Cuoco 2:29–48
Di Mario F, Jaia A (2009) Anzio. Scavi e ritrovamenti nell'archivio della Soprintendenza per i Beni Archeologici del Lazio. In: Sapelli Ragni M (ed) Anzio e Nerone. Tesori dal British Museum e dai Musei Capitolini. Gangemi Editore, Roma, pp 39–97
Diosono F (2016) Il posto degli dei: il tempio di Villa San Silvestro di Cascia e la colonizzazione romana del territorio sabino nel III secolo a.C. In: Ancillotti A, Calderini A, Massarelli R (eds) Forme e strutture della religione nell'Italia mediana antica. III Convegno Internazionale dell'Istituto di Ricerche e Documentazione sugli Antichi Umbri, Perugia-Gubbio, 21–25 settembre 2011. L'Erma di Bretschneider, Roma, pp 245–263
Diosono F (2019a) Il tempio del foro: ricostruzione architettonica e ipotesi per una lettura del monumento nel quadro della colonizzazione medio-repubblicana.- In: Battaglini G, Coarelli F, Diosono F (eds) Fregellae. Il tempio del Foro e il tempio suburbano sulla via Latina. Accademia dei Lincei, Monumenti Antichi, vol 78., serie misc. 23. Giorgio Bretschneider Editore, Roma, pp 19–28
Diosono F (2019b) Il tempio suburbano sulla via Latina e la costruzione del paesaggio sacro della colonia di Fregellae: culto e cultura materiale. In: Battaglini G, Coarelli F, Diosono F (eds) Fregellae. Il tempio del Foro e il tempio suburbano sulla via Latina, Accademia dei Lincei, Monumenti Antichi, vol 78., serie misc. 23. Giorgio Bretschneider Editore, Roma, pp 95–110
Diosono F, Braconi P, D'Angelo G, Ghini G, La Notte A (2019) Le prime fasi edilizie del tempio di Diana a Nemi. In: Cifarelli FM, Gatti S, Palombi D (eds) Oltre

"Roma medio repubblicana". Il Lazio fra i Galli e la battaglia di Zama. Atti del Convegno Internazionale, Roma, 7–9 giugno 2017. Quasar, Roma, pp 383–390

Diosono F, Fraddosio F, La Notte A, Pecere N, Piccioni MD (2022) Properties oJ foundations and podiums of roman-italic temples in Centrai Italy: the case study of Tempie B in Pietrabbondante. In: Diosono F, Le Blay F (eds) Compatangelo-Soussignan, Living with seismic phenomena in the mediterranean and reyond hetween antiquity and the middle ages, Proceedings of Cascia (25–26 October, 2019) and LeMans (2–3June, 2021) conferences. Archaeopress, Oxford, pp 379–398

Edlund Berry IEM (2008) The language of Etrusco-Italic architecture: new perspectives on Tuscan temples. Am J Archaeol 112:441–447

Finocchio G, Casablanca O, Ricciardi G, Alibrandi U, Garesci F, Chiappini M, Azzerboni B (2014) Seismic metamaterials are based on isochronous mechanical oscillators. Appl Phys Lett 104:191903

Forte M (1991) Le terrecotte ornamentali dei templi lunensi, Biblioteca di "Studi etruschi", vol 22. Olschki, Firenze

Franchetti Pardo V (2006) L'Ellenismo. In: Bozzoni C, Franchetti Pardo V, Ortolani G, Viscogliosi A (eds) L'architettura del mondo antico. Laterza, Roma-Bari, pp 117–238

Fuduli L (2015) Osservazioni sull'architettura templare della Sicilia ellenistica. Per una rilettura dei dati. Rev Archeol 60:293–345

Fusco U, Roccella V, Soriano F, Lepri B, Scavone R (2009) La stratigrafia archeologica presso il Tempio D (campagne di scavo 2005–2007). In: Mastrocinque A (ed) Grumentum romana, Atti del Convegno di studi, Moliterno, 28–29 giugno 2008. V. Porfidio, Moliterno, pp 176–216

Gan WS (2007) Gauge invariance approach to acoustic fields. In: Akiyama M (ed) Acoustical imaging, vol 29. Springer, pp 389–394

Gan WS (2018) New acoustics based on metamaterials. Springer, Singapore

Gan WS (2019) Gauge invariance approach to acoustic fields. Springer, Singapore

Ghey E (2005) Beyond the temple: blurring the boundaries of 'sacred space'. In: Bruhn J, Croxford B, Grigoropoulos D (eds) TRAC 2004 proceedings of the fourteenth annual theoretical roman archaeology conference Durham 2004. Oxbow Books, Oxford

Ghini G (2013) La terrazza inferiore: dati e recenti acquisizioni dai nuovi scavi. In: Braconi P, Coarelli F, Diosono F, Ghini G (eds) Il santuario di Diana a Nemi. Le terrazze e il ninfeo. Scavi 1989–2009, Studia archaeologica, vol 194, L'Erma di Bretschneider, Roma, pp 17–33

Ginouvès R (1992) Dictionnaire méthodique de l'architecture grecque et romaine. Tome II. Éléments constructifs. Supports, couvertures, aménagements intérieurs. Collection de l'École Française de Rome, vol 84. École française de Rome, Roma

Giuliani CF (2006) L'edilizia nell'antichità. Carocci, Roma

Gros P (1976) Les premières générations d'architectes hellénistiques à Rome. In: L'Italie préromaine et la Rome républicaine, I. Mélanges offerts à Jacques Heurgon. Publications de l'École Française de Rome, vol 27). École Française de Rome, Roma, pp 387–410

Gros P (ed) (1997) Vitruvio. De Architectura. Einaudi, Torino

Gros P (2001) L'architettura romana dagli inizi del III secolo a.C. alla fine dell'Alto Impero, Biblioteca di archeologia, vol 30. Longanesi, Milano

Gros P (2009) Les sanctuaires in summa cavea. L'enseignement des recherches récentes sur le théâtre de Pompée à Rome. In: Moretti JC (ed) Fronts de scène et lieux de culte dans le théâtre antique. Maison de l'Orient et de la Méditerranée, Lyon, pp 53–64

Guidoboni E, Comastri A, Traina G (1994) Catalog of ancient earthquakes in the Mediterranean area up to the 10th century. Istituto Nazionale di Geofisica, Roma

Guidoboni E, Ferrari G, Mariotti D, Comastri A, Tarabusi G, Sgattoni G, Valensise G (2018) CFTI5Med, Catalogo dei Forti Terremoti in Italia (461 a.C.-1997) e nell'area Mediterranea (760 a.C.-1500). Istituto Nazionale di Geofisica e - Vulcanologia-INGV. https://storing.ingv.it/cfti/cfti5

Hanson JA (1959) Roman Theater-temples. Princeton University Press, Princeton, NJ

Jaia AM (2017) Edifici di culto a Lavinium in età medio repubblicana tra continuità e innovazione. In: Caliò LM, Des Courtils J (eds) L'architettura greca in Occidente nel III secolo a.C. Atti del Convegno di Studi, Pompei-Napoli, 20–22 maggio 2015. Thiasos Monografie vol 8. Quasar, Roma, pp 265–284

Jaia AM (2019) Albano: documenti d'archivio presso l'Istituto di Topografia Antica. In: Fischetti AL, Attema PAJ (eds) Alle pendici dei Colli Albani. Dinamiche insediative e cultura materiale ai confini con Roma. University of Groningen/Groningen Institute of Archaeology and Barkhuis Publishing, Groningen, pp 185–195

Khlopotin A, Olsson P, Larsson F (2015) Transformational cloaking from seismic surface waves by micropolar metamaterials with finite couple stiffness. Wave Motion 58:53–67

La Regina A (1978) Pietrabbondante. In: Cianfarani V, Franchi Dell'Orto L, La Regina A (eds) Culture adriatiche antiche di Abruzzo e di Molise. De Luca, Roma, pp 449–489

La Regina A (1989) I Sanniti. In: Pugliese Carratelli G (ed) Italia omnium terrarum parens. La civiltà degli Enotri, Choni, Ausoni, Sanniti, Lucani, Brettii, Sicani, Siculi, Elimi, Antica madre, vol 12. Garzanti-Scheiwiller, Milano, pp 301–432

La Regina A (2014) Pietrabbondante e il Sannio antico. In: Astorri I, Di Rocco G (eds) Almanacco del Molise. I cinquant'anni della Regione Molise (1963–2013). Habacus Editore, Campobasso, pp 219–255

La Salandra V, Wenzel M, Bursi OS, Carta G, Movchan AB (2017) The conception of a 3D metamaterial-based foundation for static and seismic protection of fuel storage tanks. Front Mater 4:4–30

Lauter H (1986) L'architettura dell'Ellenismo. Biblioteca di archeologia, vol 27. Longanesi, Milano

Leonhardt U (2006) Optical conformal mapping. Science 312:5781

Linderski J (1986) The augural law. In: Haase W (ed) Religion (Heidentum: Römische Religion, Allgemeines [Forts.]) (ANRW II.16/3). De Gruyter, Berlin-Boston, pp 2146–2231

Lippolis E (2016) Il Capitolium. In Lippolis E, Osanna M (eds) I Pompeiani e i loro dei. Culti, rituali e funzioni sociali a Pompei, Atti della Giornata di Studi, Roma, 15 febbraio 2016. Scienze dell'Antichità 22.3. Quasar, Roma, pp 111–148

Livadiotti M, Rocco G (2005) Il tempio di Roma e August-o. In: Di Vita A, Livadiotti M (eds) I tre templi del lato nord-ovest del Foro Vecchio a Leptis Magna, Monografie di Archeologia Libica, vol 12. Roma, L'Erma di Bretschneider, pp 165–298

Marcattili F (2010) Bona Dea. Η ΘΕΟΣ ΓΥΝΑΙΚΑ. Archeologia Classica 61:7–40

Marcattili F (2017) Inversione della norma ed integrazione sociale: per un'interpretazione dei templi a cella trasversale. Atti della Pontificia accademia romana di archeologia. Rendiconti 89:705–744

Masturzo N (2005) Il tempio occidentale–'tempio di Liber Pater'. In: Di Vita A, Livadiotti M (eds) I tre templi del lato nord-ovest del Foro Vecchio a Leptis Magna. Monografie di Archeologia Libica, vol 12. Roma, L'Erma di Bretschneider, pp 37–161

Maurel A, Marigo JJ, Pham K, Guenneau S (2018) Surface waves from flexural and compressional resonances of beams. Phys Rev B 98:134311

Milton GW, Briane M, Willis JR (2006) On cloaking for elasticity and physical equations with a transformation invariant form. New J Phys 8:248

Montagna Pasquinucci M (1973) La decorazione architettonica del Tempio del Divo Giulio nel Foro Romano. Accademia dei Lincei, Monumenti antichi, vol 48. Accademia Nazionale dei Lincei, Roma

Mu D, Shu H, Zhao L, An S (2020) A review of research on seismic metamaterials. Adv Eng Mater 22:1901148

Muhammad G, Lim CW, Reddy JN (2019) Built-up structural steel sections as seismic metamaterials for surface wave attenuation with low-frequency wide-bandgap in layered soil medium. Eng Struct 188:440–451

Muhammad G, Tingkai W, Lim CW (2020) Forest trees as naturally available seismic metamaterials: low-frequency Rayleigh waves with extremely wide-bandgap. Int J Struct Stab Dyn 20:14. https://doi.org/10.1142/S0219455420430142

Nielsen I (1992) The Metellan Temple. In: Nielsen I, Poulsen B (eds) The Temple of Castor and Pollux. The pre-Augustan temple phases with related decorative elements, Lavori e Studi di archeologia, vol 17. De Luca, Roma, pp 87–117

Nielsen I, Poulsen B (1992) The first temple. In: Nielsen I, Poulsen B (eds) The Temple of Castor and Pollux. The pre-Augustan temple phases with related decorative elements, Lavori e Studi di archeologia, vol 17. De Luca, Roma, pp 61–79

Pendry JB, Holden AJ, Robbins DJ, Stewart WJ (1999) Magnetism from conductors and enhanced non-linear phenomena. IEEE Trans Microw Theory Tech 47(11): 2075–2984

Pensabene P (1991) Il tempio della Vittoria sul Palatino. Boll di Archeol 11–12:11–51

Pensabene P (2007) Ostiensium marmorum decus et decor, Studi architettonici, decorativi e archeometrici. Studi miscellanei, vol 33. L'Erma di Bretschneider, Roma

Potts CR (2011) The development and architectural significance of early Etrusco-Italic podia. Bulletin antieke beschaving. Annual Papers on Classical Archaeology 86:41–52

Rescigno C (2016) Il santuario di Apollo tra vecchie acquisizioni e nuove prospettive di ricerca. In: Lippolis E, Osanna M (eds) I Pompeiani e i loro dei. Culti, rituali e funzioni sociali a Pompei. Atti della Giornata di Studi, Roma., 15 febbraio 2016. Scienze dell'Antichità, vol 22.3. Quasar, Roma, pp 37–69

Rykwert J (1976) The idea of a town: the anthropology of urban form in Rome, Italy, and the ancient world. Faber and Faber, London

Scheid J (2009) Rito e Religione dei Romani. Sestante edizioni, Bergamo

Schrammen J (1906) Der Grosse Altar. Der Obere Markt (Altertümer von Pergamon 3.1). G. Reimer, Berlin

Schurig D, Mock JJ, Justice BJ, Cummer S, Pendry JB, Starr AF, Smith DR (2006) Metamaterial electromagnetic cloak at microwave frequencies. Science (NY) 314:977–980. https://doi.org/10.1126/science.1133628

Sgrosso I, Naso G (eds) (2011) Note illustrative della Carta Geologica d'Italia alla scala 1:50.000, foglio 393, Trivento. Istituto Superiore per la Protezione e la Ricerca Ambientale, Roma

Shoe Meritt LT, Edlund Berry IEM (2000) Etruscan and Republican Roman mouldings. (University Museum monograph 107). University Museum, Philadelphia

Short M (2008) Thinking places, placing thoughts: spatial metaphors of mental activity in Roman culture. I Quaderni del Ramo d'Oro on-line 1:106–129

Stamper JW (2005) The architecture of Roman temples. The republic to the middle empire. Cambridge University Press, New York/Cambridge

Strazzulla MJ (1973) Il santuario sannitico di Pietrabbondante. Tipografia Centenari, Roma

Ungureanu B, Guenneau S, Achaoui Y, Diatta A, Farhat M, Hutridurga H, Craster RV, Enoch S, Brûlé S (2019) The influence of building interactions on seismic and elastic body waves. EPJ Appl Metamaterials 6:18

Valenti M (2016) Il Capitolium e il tempio maggiore di Terracina, due esempi di podi templari a sostruzione

cava. Caratteristiche tecnico-formali, funzione e terminologia. In: Valenti M (ed) L'architettura del sacro in età romana: paesaggi, modelli, forme e comunicazione. Gangemi Editore, Roma, pp 49–62
Veselago VG (1968) The electrodynamics of substances with simultaneous negative values of ε and μ. Soviet Physics Uspekhi 10(4):509–514
Viscogliosi A (1996) Il tempio di Apollo in Circo e la formazione del linguaggio architettonico augusteo. Bullettino della Commissione Archeologica Comunale di Roma, Supplementi, 3. L'Erma di Bretschneider, Roma
von Hesberg H. (1994) Formen privater Repräsentation in der Baukunst des 2. und 1. Jahrhunderts v. Chr. Böhlau, Köln
Yanghua W (2015) The Ricker wavelet and the Lambert W function. Geophys J Int 200(1):111–115
Zink S (2012) Old and new archaeological evidence for the plan of the Palatine temple of Apollo. J Rom Archaeol 25:387–402

A Predictive Nonlinear 2D Mechanical Model for FEM Analysis of *Trait de Jupiter* Wooden Joints

8

Lorenzo Fei and Giovanni Formica

Abstract

This chapter aims to evaluate the nonlinear mechanical behaviour of the trait de jupiter for supporting through computational analysis of the use of traditional techniques to be employed in restoration interventions. This joint represents a favourite solution within traditional construction techniques due to its ability to have both compressive and tensile high strength. However, technical specifications still rely on oversimplified mechanical models, which do not account for the inherent nonlinear nature of traditional connections like the trait de jupiter, which can noticeably affect the structural response. The present work contributes to this direction by proposing a 2D nonlinear model for the contact interface of the joint. Modern case studies have been selected which preserve material evidence of construction solutions in full continuity with others discovered in archaeological contexts. Survey campaigns of the structures are then pursued on several wooden roof structures of the Roman monumental architectural heritage. As key, a nonlinear predictive mechanical model is developed and numerically analysed using the finite element method (FEM). This determines the stiffness and bearing capacity of the beam element in the function of the level of constraint offered by the joint.

Keywords

Trait de Jupiter · FEM analysis · Carpentry · Wooden truss structures · Roman monumental architectural heritage

8.1 Introduction

Timber trusses represent engineering achievements that demonstrate historically technical capabilities, design expertise, and construction abilities of great intellectual and anthropological impact. However, due to their role within buildings, roof structures are prone to deterioration and require frequent maintenance and restoration. As proved by some wooden roofing structures of the Roman monumental architectural heritage, this is particularly evident in traditional constructive solutions characterised by large tie-beams. We refer to the vast family of beams connected by *scarf joints*, widely used in historical carpentry in reason of the difficulty of finding single beams long enough to cover large spans, and of their ability to have both compressive and tensile high strength.

The case studies concern buildings constructed between the sixteenth and nineteenth centuries. Three buildings are specifically analysed: Library Hall of the Convent of Trinità dei Monti–

L. Fei (✉) · G. Formica
Dipartimento di Architettura e Restauro, Università di ROMA TRE, Rome, Italy
e-mail: lorenzo.fei@uniroma3.it

L. Pecchioli (ed.), *Archaeoseismology*, Natural Science in Archaeology,
https://doi.org/10.1007/978-3-031-28303-1_8

Table 8.1 Summaries for the case studies within the architectural complexes Convent of Trinità dei Monti

Library Hall of the Convent of Trinità dei Monti	
Place in the building	Library Hall
Truss type	Single symmetrical truss consisting of tie-beam (1), rafters (2), counter-tie-beam (1), secondary rafters (2), and posts (3)
Dimensions	Span: 11.70 m Distance between trusses: 4.40 m Rafter size (hxb): 0.34 x 0.34 m Tie-beam size (hxb): 0.35 x 0.33 m
Roof pitch angle	25.00°
Type of joint	*Trait de Jupiter* with under-squinted ends
Wood species	Chestnut
Dating	Seventeenth century; nineteenth century
Last known restoration	2000

nineteenth century (see Sect. 8.2.1, Table 8.1); Papal Basilica of St. Maria Maggiore, in the physiognomy that can be traced back to the restoration work of 1725 (see Sect. 8.2.2, Table 8.2); Salone d'Ercole in Palazzo Farnese, carried out by Giuseppe Hasslinger in 1837 (see Sect. 8.2.3, Table 8.3). Nonetheless, similar solutions are well known from archaeological contexts (Lipke 1984; Adam 1988; Linder et al. 2003; McGrail 2009; Pomey et al. 2005; Beltrame 2016). In particular, maritime findings allow us to state how ancient naval architecture should be understood as the archetype of wooden architecture, having adopted sophisticated technical solutions (Beltrame 2016), used today in building practices.

Our methodology first involved the investigation of literary sources, starting with Vitruvius' *De Architectura Libri Decem* and its subsequent interpreters and technical manuals, and then the study of examples selected by typological eloquence. Afterwards, surveys of the structures were conducted, and graphical representations are elaborated. The analysis of the collected data highlights the problems related to the connection of the wooden elements for long-span tie-beams, which are known to be identified as weak points of the system.

Such connections were commonly adopted in Roman historical buildings as favourite construction solutions, also in reason of the seismic nature of the city of Rome, where earthquakes reaching

Table 8.2 Information about the Basilica of St. Maria Maggiore

Papal Basilica of St. Maria Maggiore	
Place in the building	Central nave
Truss type	Double symmetrical truss each one consisting of tie-beam (1), rafters (2), counter-tie-beam (1), secondary rafters (2), posts (3), and tie-beam tensioner (1)
Dimensions	Span: 15.50 m Distance between trusses: 3.50 m Rafter size (hxb): 0.34 x 0.27 m Tie-beam size (hxb): 0.34 x 0.34 m
Roof pitch angle	26.50°
Type of joint	Double straight-splayed *scarf joint*
Wood species	Chestnut
Dating	1437, 1720,1724, 1725, 1900, 1975
Last known restoration	1989–1991

Table 8.3 Summary for the case study within the architectural complex of and of the Palazzo Farnese in Rome

Salone d'Ercole at Palazzo Farnese	
Place in the building	Salone d'Ercole, first floor
Truss type	Superimposed single symmetrical trusses. The first from above is composed of tie-beam (1), rafters (2), counter-tie-beam (1), secondary rafters (2), posts (3). The second from above is composed of tie-beam (1), rafters (2), secondary rafters (2), braces (2), posts (3)
Dimensions (upper truss)	Span: 14.36 m Distance between trusses: 4.80 m Rafter size (hxb): 0.35 x 0.27 m Tie-beam size (hxb): 0.40 x 0.27 m
Roof pitch angle	24.00°
Type of joint	Double stop-splayed *scarf joint*
Wood species	–
Dating	Sixteenth century
Last known restoration	–

up to grade VIII on the Mercalli–Cancani–Sieberg scale (MCS-1930) were measured in the past (Molin et al. 1995; Donati et al. 2008; Fei et al. 2020). These events may cause collapse mechanisms due to the lack of connections, making local buckling of roof frames possible (Giuffrè et al. 1988; Cangi 2012).

Finally, a two-dimensional nonlinear mechanical model will be proposed, capable of providing a numerical evaluation of the stiffness and load-bearing capacity of the beam elements using Finite Element simulations. Such predictions are intended to contribute to a first insight on the constraint level for the *Trait de Jupiter,* accounting for the actual nonlinear nature of the wooden connection. In accordance with the objectives, procedures and criteria established in the UNI 11119:2004 standard for the assessment of the performance of in-situ wood elements in load-bearing structures of buildings included in the cultural heritage area, through the use of non-destructive analysis methods.

8.2 Case Studies

In the following section, we summarise aforementioned case studies presenting three different types of *scarf joints*. They refer to particularly complex structures resulting from consolidation and restoration works carried out over time: such structures cover large spans, and therefore are subjected to demanding operating conditions.

A graphic representation of the key architectural elements of all the case studies is reported in Fig. 8.1.

8.2.1 Library of the Convent of Trinità dei Monti

The first case study has been detected in the second truss of the roof structure of the Library Hall of the Convent of Trinità dei Monti [1] (Fig. 8.1a). The truss under examination is the result of a nineteenth-century consolidation of the simpler *king post truss*. In this case, a structural trestle—consisting of two secondary rafters, a counter tie-beam, and two posts—was added to the former truss. Although the width of the hall is 11.70 m, this solution allows spans of up to 20 m (Mazzocchi 1871).

Compared to the usual layout, the *stop-splayed scarf joint* tabled with under-squinted abutments presents an anomaly, being shaped on the vertical faces of the tie-beam instead of the horizontal ones.

[1] Special thanks to the *Service des Travaux et Bâtiments français à Rome,*Laurent Delfour (Architecte en chef du STBI), and Agnès Chodzko (Architecte chargée d'opération). We also acknowledge the cooperation of the Archives des Pieux Etablissements de la France à Rome et Lorette, especially Dr. Michela Berti, who made the documents available for study.

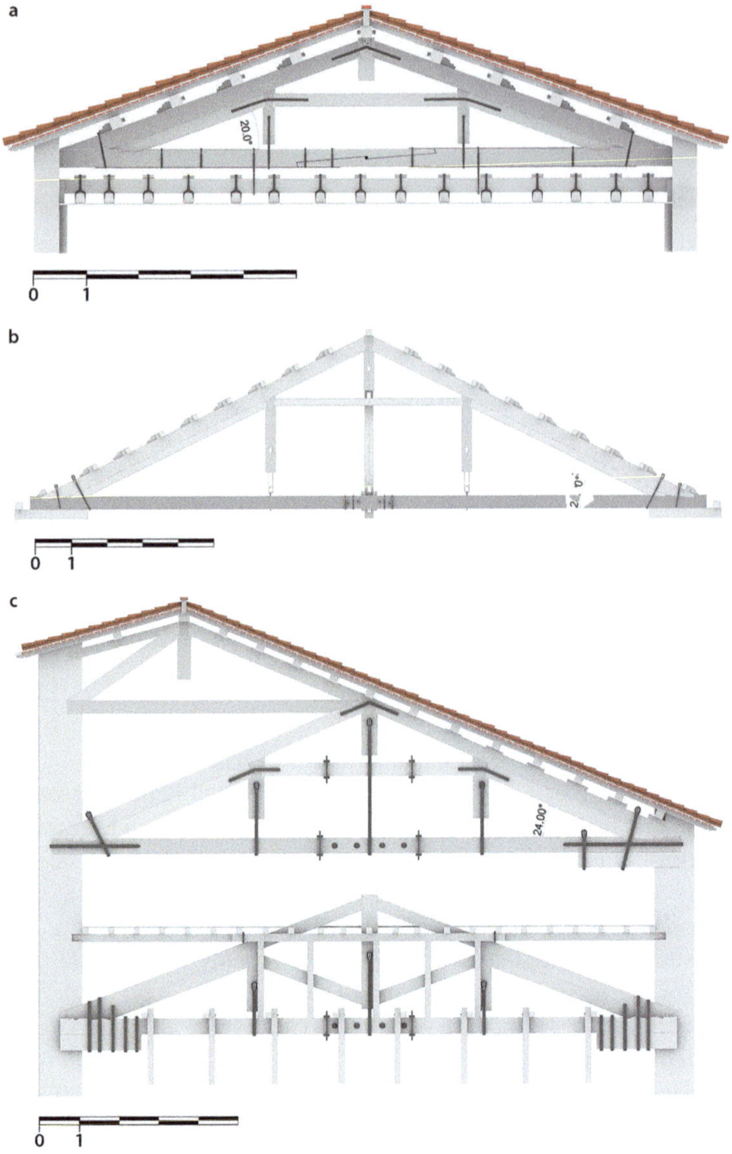

Fig. 8.1 (**a**) Front elevation of the second truss of the roof structure of the Library Hall of the Convent of Trinità dei Monti; total span 11.70 m (survey and graphic representation by L. Fei, 2020); (**b**) Front elevation of the double truss structure of the Basilica of St. Maria Maggiore, central nave; span between bracket supports 15.50 m, total span 18.60 m (survey and graphic representation by Lorenzo Fei, 2020); (**c**) Palazzo Farnese, Salone d'Ercole. The framed structure from Giuseppe Hasslinger's survey (graphic representation by Lorenzo Fei, 2020)

8.2.2 Papal Basilica of St. Maria Maggiore

The second case study was observed in the fourth pair of trusses of the central nave of the Papal Basilica of St. Maria Maggiore (Fig. 8.1b). [2] The original structure dates to the fifteenth century, but its current physiognomy is the result of a consolidation work carried out in the eighteenth century, when the lateral posts were added, and in the twentieth century when the structure was extensively recomposed and the ironwork replaced. The span of the hall, considering the cantilevers, is 15.50 m, but this type of truss is adequate for spans up to 24 m (Mazzocchi 1871).

[2] We wish to express our utmost gratitude to his Eminence and Most Reverend Stanisław Cardinal Ryłko, Archpriest of the Papal Basilica of Santa Maria Maggiore and to Don Marco Cocuzza, Most Reverend Coadjutor of the Chapter for having granted us access to the attic rooms of the Basilica. We also thank Dr. Pier Giorgio Cataldi, archivist at the Archivio della Basilica Papale di Santa Maria Maggiore (ABPSMM) for his competence and availability.

The *double straight-splayed scarf joint* under examination is a rather rudimentary example concerning the previous type of *Trait de Jupiter*, which justifies consolidation works of reinforcement of the truss with bolts, clamping, and support brackets connected to iron ankles placed at the base of the tie-beam tensioner.

8.2.3 Salone d'Ercole at Palazzo Farnese

The third case was modelled from the survey done by Giuseppe Hasslinger (1837) of the trusses of the Salone d'Ercole in Palazzo Farnese, demolished in the early twentieth century (Fig. 8.1c). [3] The original structure dates to the sixteenth century. It consists of two superimposed trusses, the former supporting the roof and the latter the ceiling covering the Salone. The extent of the hall is 14.36 m, although both types can be used for spans up to 20 m (Mazzocchi 1871).

The examined *double stop-splayed scarf joint* belonging to the upper truss is a simplification of the canonical *Trait de Jupiter*, requiring the use of abundant ironwork to tighten the joint. In this case, as in the previous one, the layout of the joints is consistent with the requirements of the so-called 'rule of art'.

8.3 Mechanical Predictive Analysis

The mechanical behaviour of the *Trait de Jupiter* joints is investigated by developing a nonlinear two-dimensional mechanical model for a numerical prediction via Finite Element simulations of both stiffness and bearing capacity of the beam element. The structural performance is pursued by evaluating the influence of the connection on the behaviour of the whole beam. Such an evaluation allows us to give some preliminary comments on the actual level of constraint offered by the *scarf joint.*

We perform in the most demanding conditions, i.e., assuming a dry connection for the joint, which is therefore modelled without brackets and pre-tensioning wedges; when considering such reinforcements, the beam shows actually a largely linear response under varying loading conditions, as indicated in technical specifications.

8.3.1 Development of the Model

A two-dimensional model is developed and implemented in COMSOL Multiphysics. In order to keep the focus on the influence of *scarf joints* on the structure response, some simplified assumptions are adopted: linear-elastic behaviour is prescribed for the (wood) material, instead of the more properly viscoelastic behaviour (Konopka et al. 2017): in particular, accounting for its orthotropic nature, an equivalent elastic material is employed (Kretschmann 2010). Finally, we do not consider any fatigue limitation on the stress state related to hygroscopic shrinkage or swelling (Bergman 2010; Abdelmohsen et al. 2019).

The resulting governing equations are, therefore, those of a geometrically nonlinear elastic problem based on the second Piola–Kirchhoff stress.

The key modelling contribution is the contact law resolving the unilateral structural response at the contact surface. This is a highly nonlinear problem that we handle with the approach of the contact pair: a pair is defined on the joint boundaries (simulated as separate faces) where the parts can come into contact, but they cannot penetrate each other under deformation. Such an approach allows us to model the contact surface as an interface element characterised by a cohesive compression friction law and a softening tensile response.

The stress state along the contact surface is therefore bounded by a yielding criterion that obeys to the following condition:

$$T_t \leq \min(\mu T_n + T_{\text{cohe}}, T_{t,\max}) \text{with } \mu = \mu_{\text{stat}}$$

[3] Special thanks to the architects Pierre-Antoine Gatier (Architecte en chef des Monuments Historiques of the departments of Paris and Rome) and Giulia Gabellini for sharing the documents stored at the Akademie der Bildenden Künste Wien Kupferstichkabinett.

where μ_{stat} is the static frictional coefficient, T_{cohe} is the cohesion sliding resistance that supplies traction stress which must be overcome before sliding can occur, $T_{t,\,max}$ is the maximum tangential traction. When the tangential traction exceeds this value, a slip occurs, independently of the pressure level.

8.3.2 Geometry and Material Characteristics

As for the geometry, modelled with CAD software, the tie-beam combined with *Trait de Jupiter* of the Trinità dei Monti's library is an example. The dimensional parameters are as follows:

- Height: 35 cm.
- Thickness: 33 cm.
- Length: 760 cm.

The joint is built of chestnut wood with the following material characteristics (UNI EN 11035-2:2010):

- Density: 580 kg/m3
- Young's modulus: 12,500 MPa
- Poisson's coefficient: 0.3
- s_{id}: 12 MPa

8.3.3 Method of Continuation Analysis

Pseudo-temporal loading conditions in strain-driven mode are imposed in order to conduct a step-by-step analysis with an increasing displacement load pattern. The prescribed displacement increments have to be considered as fictitious time steps, where inertial contributions are neglected, allowing the analysis to proceed with stationary operators.

For each increment of a scalar continuation parameter governing the loading history, we solve incrementally the steady-state equilibrium conditions via an iterative error-minimisation scheme.

8.4 Numerical Results

8.4.1 Four-Point Test for the Evaluation of Beam Element Stiffness and Load-Bearing Capacity

The first set of analyses is conducted to determine the effective elastic response of the *Trait de Jupiter* joint compared with the other two types of joints, namely d*ouble straight-splayed scarf joint* and the d*ouble stop-splayed scarf joint*. The numerical analyses, simulating a four-point bending test, are driven by prescribing a displacement load history on two points, symmetrically arranged with respect to the midline and spaced one and a half times the height of the beam (52.5 cm) concerning the joint (see Fig. 8.2); the displacement history is driven by a scalar parameter (u) playing the role of continuation parameter. On the other hand, constraints of zero displacements are imposed on the two ends of the beam. The results are compared with an ideal continuous beam, obtained by assuming a perfect connection at the joint interface, and characterised by the same dimensions and material; such a reference solution exhibits only a linear response.

The results reported in Fig. 8.3 demonstrate that for a maximum level of deflection reached at 10% of the beam length, 7.60 m, the beam with *Trait de Jupiter* joint performs better than the other two beams showing only a 35% reduction of the stress, to the ideal perfectly linear-elastic case. This high level of deformation also reveals information on the load-bearing capacity of the connected beam: the beam using the d*ouble straight-splayed scarf joint* appears to be still in a linear regime, however, showing higher deformability (about four times concerning the ideal case). The other two beams seem to reach a sort of plateau in the stress–strain curve, which, in the absence of a material strength prescription, i.e., a perfectly elastic material, can be ascribed only to the nonlinear nature of the joint reactive response. In this sense, the load-bearing capacity of the beam with *Trait de Jupiter* joint is greater

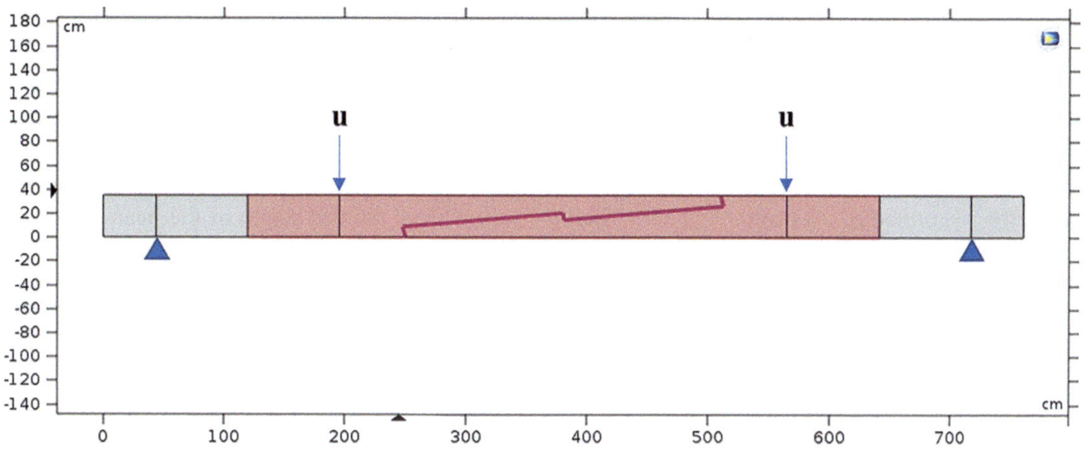

Fig. 8.2 Diagram of the displacements and constraints imposed on the beam. The highlighted area indicates the Von Mises stress integration domain

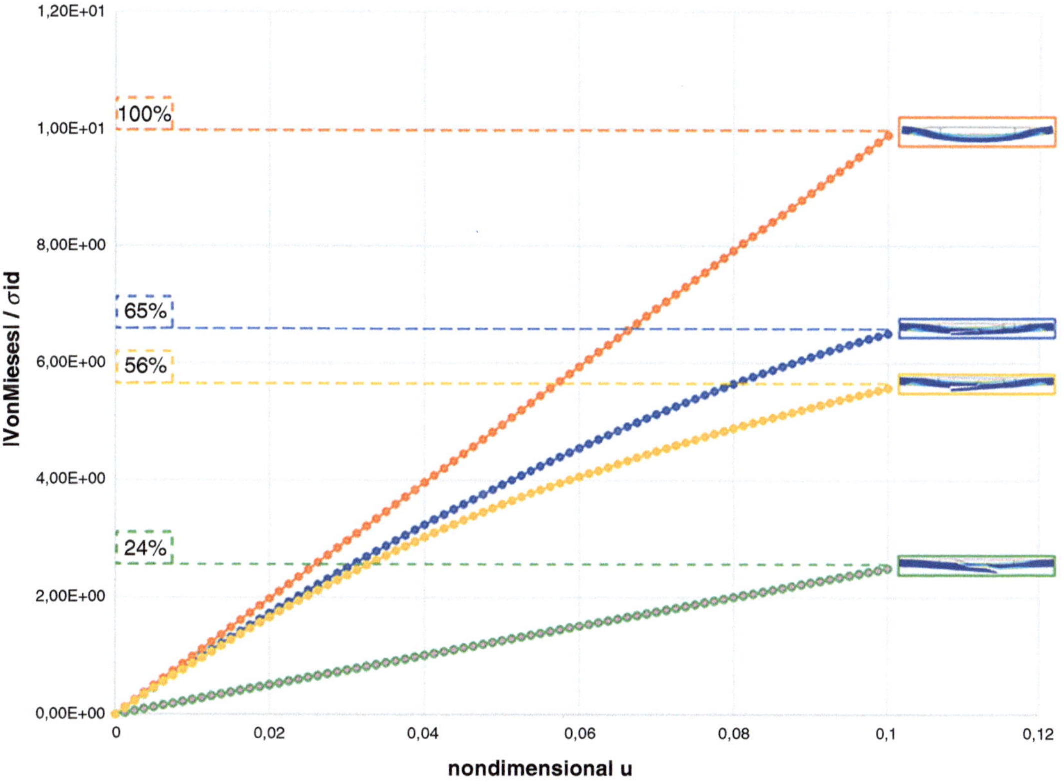

Fig. 8.3 Nondimensional force–displacement curves for different joint connections: *Double straight-splayed scarf joint* (in green lines), *Double stop-splayed scarf joint* (yellow), *Trait de Jupiter* with under-squinted ends (blue), perfectly continuous joint (red). Forces are the averaged Von Mises stress divided by an ideal linear-elastic stress; the prescribed displacement u is normalised with respect to the beam length

than about 10% than the one with *double stop-splayed scarf* joint.

Under working conditions (deflection at 1% of the beam length), the curves of these two latter beams are both very close to the ideal one, meaning that they can be designed as a continuous, perfectly connected beam. The same considerations can be done for the effective stiffness of such two jointed beams, which turns out to be just 12% lower than that of a perfectly continuous beam.

8.4.2 A First Insight on the Level of Constraint for the *Trait de Jupiter*

The second set of analyses is conducted to discuss on the degree of constraint offered with the *Trait de Jupiter* joint by forcing the jointed beam to undergo different deformed configurations, namely axial, flexural, and shear strain. Such deformations are obtained by imposing suitable displacement conditions, parametrised in terms of a scalar u, and prescribed on the extremal vertical boundaries at left and right end cross-sections. All other characteristics of the previous test do not vary, except for the length of the beam, which is reduced to 3.70 m to better emphasise the joint response in its immediate surroundings. Each deformation test is performed twice, according to the two possible loading directions.

Stress–strain curves are carried out and reported in Figs.8.4, 8.5, and 8.6. The most critical test turns out to be the flexural one, see Fig. 8.4, where we also reported the corresponding deformed configurations at the final stage of the analysis. The geometry of the dry joint causes an unstable behaviour, governed by a sudden opening of the joint, leading to large deflections with loss load-bearing capacity. This phenomenon is particularly evident in the case of

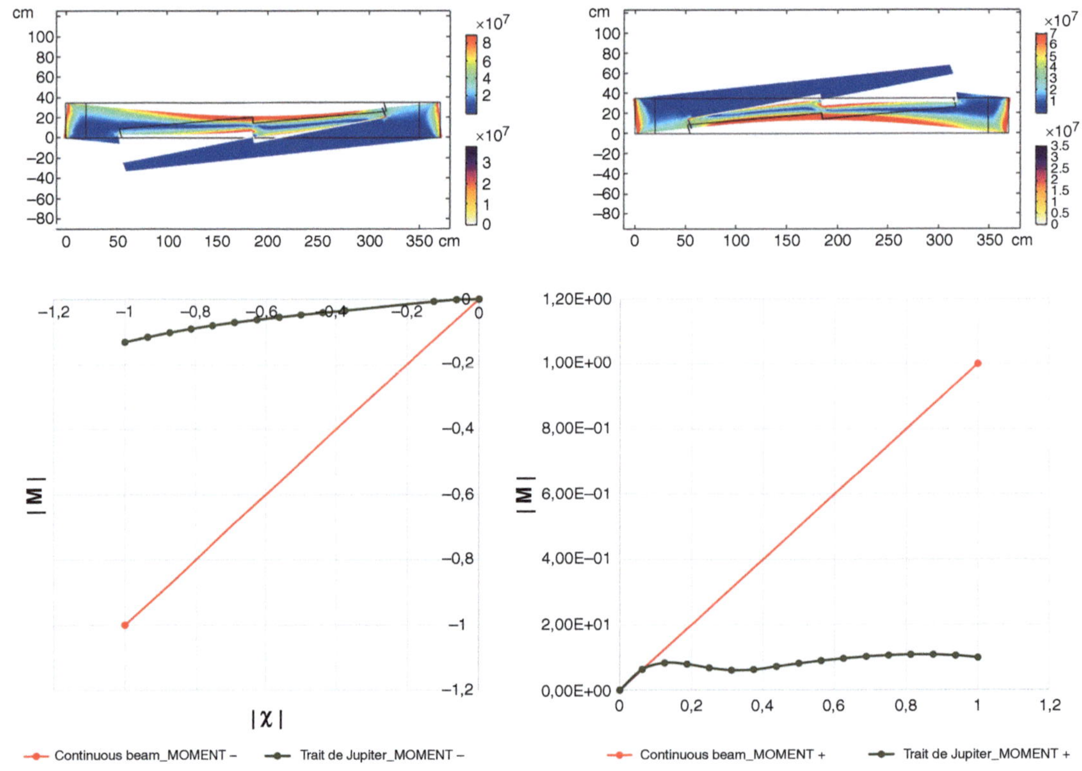

Fig. 8.4 Flexural test. Deformed configurations (on the top), bending nondimensional stress–strain curves for both loading directions (on the bottom)

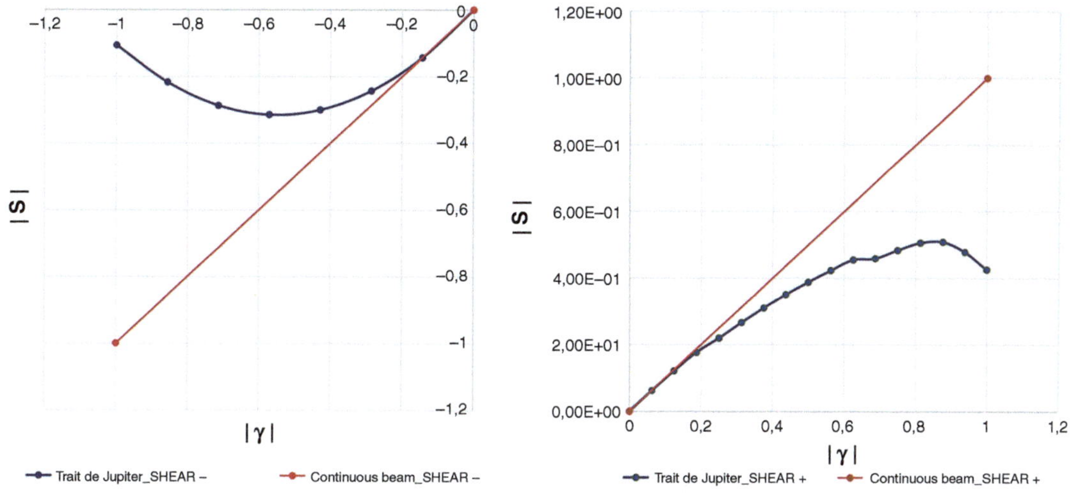

Fig. 8.5 Shear test. Resulting shear nondimensional stress–strain curves for both loading directions

intrados flexure where, as expected, the opening of the joint occurs for a low loading level at the beginning of the analysis; in this case, which leads the structural response curve moves further away from the ideal one.

In the other deformation cases, a loss in load-bearing capacity is also obtained. The shear stress–strain curves in Fig. 8.5 show limit stress followed by a softening response; the achieved level of peak stress suggests reducing at least 50% of the load-bearing capacity under shear conditions. Similar results are obtained in the tensile test (Fig. 8.6b). At the same time, in the compressive loading condition, the joint exhibits a perfect linear response, up to the stress level where buckling effects can occur.

8.5 Concluding Remarks and Future Developments

In the present work, we proved how the *Trait de Jupiter* joint can be regarded as an efficient hinged node, although requiring suitable

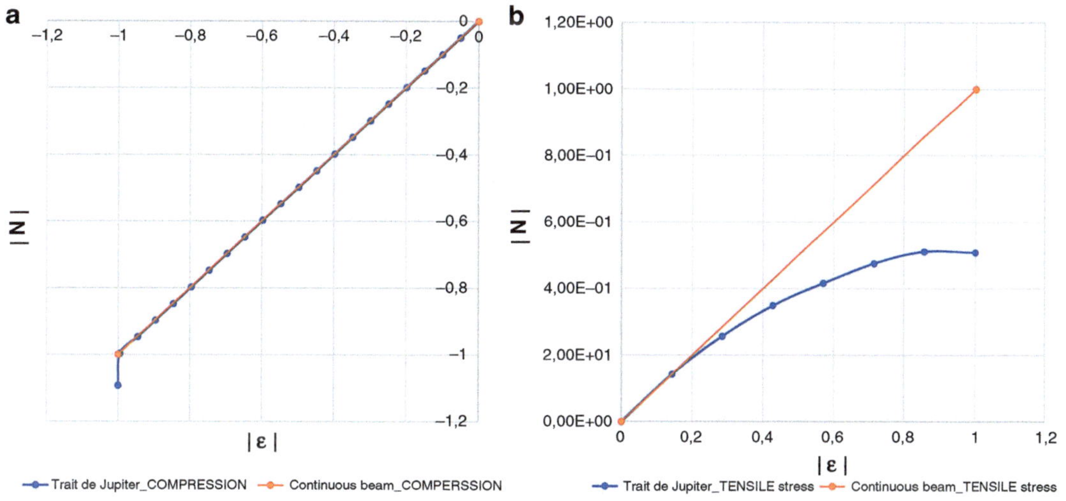

Fig. 8.6 Axial test. Resulting axial nondimensional stress–strain curves for both loading directions

limitations on both force and displacement to be accounted for in a proper design. Such limitations were successfully estimated through the proposed nonlinear model of the joint, regarded as a contact cohesive interface.

Through comparative analyses where a four-point test is simulated, the beams with *Trait de Jupiter* joint exhibit a quite high-bending response, with a reduction of only 35% on the maximum stress.

Successive analyses, addressed to compute the beam's structural response in different loading conditions, reveal how such a load-bearing capacity has to be suitably and furthermore reduced. The flexural response clearly indicates that the *Trait de Jupiter* jointed beam can just admit low levels of bending stress. In the other loading conditions, the jointed beam seems to yield stress responses acceptable for design purposes; however, in both axial and shear deformations, corrective terms have to be explored to model the beam response properly.

The conducted analyses have therefore to be finally regarded as benchmark tests to drive future developments of enhanced beam models, able to frame the joint nonlinear response directly in the beam structural response, and thus paving the way to analyses of complex wood beams frames.

References

Abdelmohsen S, Adriaenssens S, El-Dabaa R, Gabriele S, Olivieri L, Teresi L (2019) A multi physics approach for modeling hygroscopic behavior in wood low-tech architectural adaptive systems. Comput Aided Des 106:43–53. https://doi.org/10.1016/j.cad.2018.07.005

Adam JP (1988) L'arte di costruire presso i Romani. Materiali e ricerche. Longanesi, Milano

Beltrame C (2016) Archeologia marittima del Mediterraneo. Navi, merci e porti dall'antichità all'età moderna. Carocci editore, Roma

Bergman R (2010) Drying and control of moisture content and dimensional changes. In: Wood handbook–wood as an engineering material. In: General technical report FPL-GTR-190, Laboratory. Madison, WI: Laboratory, U.S. Department of Agriculture–Forest Service–Forest Products, pp 13.13–5.20

Cangi G (2012) Manuale del recupero strutturale e antisismico, 2nd edn. Dei, Roma

Donati S, Cifelli F, Funiciello F (2008) Indagini macrosismiche ad alta densità per lo studio del risentimento sismico nella città di Roma. La geologia di Roma: dal Centro Storico alla Periferia. Memorie Descrittive della Carta Geologica d'Italia LXXX:13–30

Fei L, Angelucci F, Pugliano A (2020) Edilizia storica romana: cartografia dei danni in scala MCS causati dai terremoti storici. Strumento critico per la valutazione della vulnerabilità sismica. In: Minutoli G (ed) ReUSO 2020. Restauro: temi contemporanei per un confronto dialettico. Didacomunicationlab, Firenze, pp 714–723

Giuffrè A, Zampilli M, Ceradini V, Jacovoni F, Pugliano A (1988) Centri Storici in zona sismica. Analisi tipologica della danneggi abilità e tecniche d'intervento conservativo, Castelvetere sul Calore. Cer-Edilstampa, Roma

Konopka D, Gebhardt C, Kaliske M (2017) Numerical modelling of wooden structures. J Cult Herit 27S:93–102. https://doi.org/10.1016/j.culher.2015.09.008

Kretschmann D E (2010) Mechanical properties of wood. In: Wood handbook–wood as an engineering material. In: General technical report FPL-GTR-190, Laboratory. Madison, WI: Laboratory, U.-S. Department of Agriculture–Forest Service–Forest Products, pp 5.1–5.46

Linder E, Kahanov Y (2003) The Ma'agan Mikhael ship, The recovery of a 2400-year-old merchantman final report, vol I. Israel Exploration Society, Haifa

Lipke P (1984) The Royal Ship of Cheops. BAR International Series, Oxford

Mazzocchi L (1871) Trattato sulle costruzioni in legno. Vallardi, Milano

McGrail S (2009) Sea transport, part 1: Ships and navigation. In: Oleson JP (ed) The Oxford handbook of engineering and technology in the classical world. Oxford University Press, Oxford, pp 606–637

Molin D, Castanetto S, Di Loreto E, Eugenio G, Liperi L, Narcisi B, Paciello A, Riguzzi F, Rossi A, Tertulliani A, Traina G (1995) Sismicità. In: Funiciello R (ed) Memorie descrittive della carta geologica d'Italia, La geologia di Roma. Il Centro Storico, vol I. Istituto Poligrafico e Zecca dello Stato, Roma, pp 323–408

Pomey P, Rieth E, Rival M (2005) L'archéologie navale. Errance, Paris

UNI 11119 (2004) Beni culturali–Manufatti lignei Strutture portanti degli edifici–Ispezione in situ per la diagnosi degli elementi in opera

UNI EN 11035-2 (2010) Legno strutturale–Classificazione a vista dei legnami secondo la resistenza meccanica–Parte 2: Regole per la classificazione a vista secondo la resistenza meccanica e valori caratteristici per tipi di legname strutturale

Part IV

Historical Seismicity

Roman Corinth and Wider Area (Greece): An Area Characterized by Debates on Ancient Earthquakes and by Earthquakes with Observable Signs

9

Stathis C. Stiros

Abstract

The wider Corinth region is characterized by intense seismicity which has influenced archaeological investigations for more than a century. For this reason, the archaeological stratigraphy of its Roman period is mostly described in terms of three destructive earthquakes in the second half of the first, fourth, and sixth centuries, correlated or inferred from events poorly mentioned in ancient texts. Mainly because this association is dubious, there have been proposed alternative interpretations, assuming destructions during invasions or during the Christianization of the region. Hence, Roman Corinth covers the whole range of ideas that can be proposed to explain archaeological evolution in a seismically active region.

In addition, signs of earthquakes are not limited to evidence essentially lost due to the progress of an excavation or after the restoration of a monument. In fact, signs of earthquakes, mostly in the form of coastal changes, are clear and easily observable: an at least two-phase subsidence of Kenchreai, the eastern harbour of Corinth, in which about 100 packs of abandoned unique glass posters of *opus sectile* testify to a major catastrophe; slight uplift of Lechaion, the western harbour of Corinth, excavated in a sandy environment, and a 4 m uplift of the harbour of Aigeira are among the signs of earthquakes along the south coast of the Gulf of Corinth. Coastal uplifts and subsidence are also observed along the northern coast. At the same time, the debate on the causes of the destruction of a major temple by earthquakes or by Christian zealots is summarized in the case of the Temple of Nemea, in the hinterland of Corinth.

Keywords

Archaeoseismology · Earthquake · Seismic coastal uplift/subsidence

9.1 Introduction

In 1926 a very strong earthquake in the Aegean Arc affected Rhodes and Crete, disrupted the excavations of Sir Arthur Evans at Knossos in Crete and allowed him to propose his theory that earthquakes had played a role in the collapse of the Minoan palaces (Evans 1928).

It was not the first time that earthquakes were assumed to have played a key role in the history of ancient sites. However, the case of Knossos was spectacular: the pattern of seismic damage in modern buildings was recognized in ancient

S. C. Stiros (✉)
Department of Civil Engineering (Τμήμα Πολιτικών Μηχανικών), Patras University (Πανεπιστήμιο Πατρών), Patras, Greece
e-mail: stiros@upatras.gr

L. Pecchioli (ed.), *Archaeoseismology*, Natural Science in Archaeology,
https://doi.org/10.1007/978-3-031-28303-1_9

remains during an ongoing excavation. This, in combination with the authority of Evans and the importance of Knossos, led numerous later archaeologists and historians to uncritically adopt the earthquake scenario to explain destructions in ancient buildings and stratigraphic discontinuities in various ancient sites in different periods (see Rapp 1987; Stiros 1996). In fact, in some of these cases, the signs of earthquakes were rather clear, but in many other cases, the evidence of earthquakes was poor, or there was no evidence. Further, earthquakes were regarded as a *Deus ex machina* to explain the destruction of layers or abandonment of buildings and whole sites (Stiros 1996) or even entire civilizations (Rapp 1987).

In contrast to these a priori earthquake scenarios, other investigators, mostly historians, preferred a priori anthropogenic scenarios for similar effects. Inspired by the reported generalized destruction of Athens, Thebes, or Carthage by the Persian, the Macedonian, or the Roman Army, respectively, some investigators preferred to ignore any role of earthquakes and tried to assign observed destructions and abandonment of sites to anthropogenic events, to reported or inferred invasions or to political, religious, and social unrest. Hence, prehistoric soldiers, camel-riding Bedouins, Christian zealots, or invading Goths were regarded as responsible for the destruction of temples, houses, or whole towns. In some cases, the observed destruction was at scales beyond the limits of human forces in periods in which no explosives had been invented (for references, see Di Vita 1995; Stiros 1996), while in some periods, historical reports for the destruction of a pagan temple in Asia have been regarded as a parallel for the collapse of buildings in ancient Olympia; Alexandris et al. (2014)).

In response to these contrasting views, and based on inspiration and evidence from previous investigators (Karcz and Kafri 1978; Rapp 1987), the interdisciplinary field of Archaeoseismology emerged in the 1990s, based on interdisciplinary investigations in various parts of the world, but with emphasis on remains in the ancient Greek and Roman World (cf. Stiros 1996; Nur and Ron 1997; Ellenblum et al. 1998; Nikonov 1998; Hancock and Altunel 1997; Meghraoui et al. 2003; Galadini et al. 2006; Korjenkov et al. 2006; Karakhanian et al. 2008; Rodríguez-Pascua et al. 2011; Jusseret et al. 2013; Kázmér and Mayor 2015).

As a result, three different approaches/scenarios dominate the reconstructions of the history of some monuments which are characterized by deformations, repairs or abandonment (or even ancient towns and regions in which they belong): (i) earthquakes are regarded as *Deus ex machina* to explain these effects; (ii) these effects are a priori assigned to anthropogenic factors and (iii) are the focus of archaeoseismic (or archaeoseismological) analyses.

Three reasons are responsible for this. First, archaeological evidence is usually fragmentary. Second, evidence is usually lost after an excavation or a restoration. For example, stratigraphic details are lost with the progress of an excavation, and structural deformation in columns or any other structure is obliterated after restorations (Fig. 9.1), or after the most important findings are moved to a museum. Hence, the signs of a possible earthquake are obliterated, and any modern scenarios for destruction, deformation, or abandonment of a structure can only rely on some old photos, some notes in excavation diaries, and occasionally only on detailed observations and analyses. Third, different processes can produce similar effects; for example an earthquake and a landslide. On this point focused the criticism of Karcz and Kafri (1978), and on these grounds, Stiros (1996) suggested that any archaeoseismological result can ONLY be probabilistic, with the probability increasing with the available lines of evidence.

For all these reasons, although evidence of ancient earthquakes has been claimed for various sites, in fact, well-documented and clearly observable signs are available for very few cases only. A question arising is whether there exists an area which is seismically active (so that any earthquake scenario can be a priori convincing), with numerous signs 'archaeoseismological' features

Fig. 9.1 Aerial view of the Nemea Temple, showing collapsed columns in an arrangement usually reminiscent of seismic collapse. Three columns surviving since antiquity are marked with red arrows. Several other columns have been restored, and the collapse pattern is no longer visible. Modified after Makris and Psychogios (2004). For details, cf. Stiros (2020c)

(exhibits), with ideas and models can be tested and evaluated, and which can be used to train younger generation investigators.

The wider Corinth region, especially in its Roman phase, is a rare case in which both ancient earthquakes and anthropogenic effects have been claimed to constrain its archaeological stratigraphy, and where signs of several earthquakes are clear. Roman Corinth hence represents an extraordinary paradigm, and for this reason, is the focus of this article. In fact, this area is close to Athens, and the sites discussed below can be covered through a one or two-days only-long visit.

The term 'Corinth region' in this article refers not only to the territory of the ancient Greek 'polis' (town state) and of Roman Corinth, but to a wider region, which covers several ritual centres and other nearby ancient towns which are at a distance of a few dozen kilometres from the centre of ancient Corinth.

9.2 Roman Corinth and Its Surroundings

Corinth was a famous ancient Greek town with colonies in the Ionian Islands, in the Adriatic, in Sicily, and in the North Aegean. Due to its location next to the Isthmus, a narrow zone of land separating and connecting the Aegean Sea and the Gulf of Corinth and the Ionian Sea, it became a hub in transportations between the Aegean and the Eastern Mediterranean, on the one hand, and the Ionian/Adriatic Sea and the Western Mediterranean on the other (Fig. 9.2). This role was based on two major harbours, Lechaion to the west, in the Gulf of Corinth, and Kenchreai to the east, in the Saronic Gulf, as well as to Diolkos, a paved corridor, 6 km long, crossing the Isthmus, roughly at the path of the present-day Canal, permitting ships to enter one sea to the other (Fig. 9.3). This strategic position gave this town great wealth, reflected in the famous classical Corinthian pottery and a famous proverb Strabo mentioned in his *Geographia* (*VIII, 6, 20*) 'Not for every man is the voyage to Corinth'.

The ancient Greek town of Corinth was destroyed as a political entity by the Roman Army in 146 BC. It was partly deserted (Gebhard and Dickie 2003), but Corinth was re-founded as a Roman colony in 44 BC (*Colonia Laus Iulia Corinthiensis*). By the middle of the first century, it became one of the most important trade and cultural centres in the wider region. However, since the second half of the fourth century AD, the town was in decline, with intervals of reconstruction of unknown extent.

Fig. 9.2 Corinth represented a crossroad and a transportation hub between Eastern Mediterranean and Italy. Arrows indicate a sense of marine transport from east to west. The Gulf of Corinth and Saronic Gulf lie, respectively, to the west and east of Corinth. After Stiros (2020a)

Important towns and major ritual centres marked the wider Corinth area: the Hera temple in the Perachora Peninsula, in a coastal site on the Greece mainland; the temple of Zeus at Nemea, in the Corinth hinterland, where the Nemean Games took place; and, the Poseidon Temple at the Isthmus of Corinth, where the Isthmian Games took place. Next to the harbour of Lechaion, the

Fig. 9.3 Location map. Two subparallel lines at Ag. Vassilios indicate the approximate traces of normal surface faults which caused subsidence of the area around them during the 1981 earthquakes. A bold line marks modern Canal in the Isthmus of Corinth, roughly following ancient *Diolkos*

Leonides Basilica, one of the largest basilicas in early Christian times, was built.

The wider Corinth region has attracted archaeological studies and excavations since the nineteenth century. These studies have brought to light numerous ancient remains, in which various destruction layers and periods of abandonment were recognized and fed theories for the evolution of the ancient town and the wider region.

9.3 The Three Scenarios for the Corinth Area

In this section, the scenario of certain historical earthquakes assumed to explain the stratigraphic history of the Corinth area and the alternative scenario of anthropogenic effects is examined. An inventory of the main cases of observable effects of ancient earthquakes is examined in the following paragraphs.

9.3.1 The Scenario of Destructive Earthquakes

The wider Corinth area is well-known for its seismicity and seismic faulting, especially after the benchmark 1981 earthquakes (Jackson et al. 1982; Fig. 9.2), and for coastal signs of ancient earthquakes (e.g. Pirazzoli et al. 1994). In the last two centuries, several earthquakes have produced major destruction in this area; among them, the 1928 and 1981 Corinth earthquakes, which inspired Greece's first and second antiseismic building Codes (Papazachos and Papazachou 1997).

For this reason, since the first days of systematic archaeological studies and excavations in the wider Corinth area, investigators have had first-hand experience of earthquakes, especially those of 1858 and 1928, and their impacts on the town infrastructure and the population. Consequently, archaeologists suspected that observed or inferred stratigraphic discontinuities are likely to reflect earthquake effects.

However, in the first decades of the twentieth century, no possibility to substantiate such findings existed, nor specific information for seismic damage to Corinth was discovered in ancient sources. For this reason, three discontinuities dominating the archaeological stratigraphy of the town (destruction layers or abandonment periods) have been explained as earthquake effects, and they have been assigned to earthquakes known from historical sources. But in the absence of specific information for earthquakes in Corinth, they have been correlated with earthquakes reported for other parts of Greece in 77 AD, 365 AD, 375 AD, and 551 AD (Guidoboni et al. 1994). This association is found in early (Finley 1932) and more recent (Ivison 1996; Avramea 1997) studies for Corinth, and these earthquakes are regarded as benchmarks in the Corinthian archaeological stratigraphy and palaeogeography (Ibrahim et al. 1976). This association is broadly adopted even by modern excavators (e.g. Williams II and Zervos 1985), but not without questioning (e.g. Sanders 1999).

In fact, among all these events, only for the 77 AD event, there exists some reference to Corinth by a sixth century source (*Malalas 261*), and this event remains unclear. The 375 earthquake is a duplicate of the 365 AD Crete earthquake, which is not expected to have any specific impacts on Corinth (Stiros 2020b). The 551 earthquake is probably an amalgamation of different earthquakes which hit central Greece, but in a region at least 100 km away from Corinth. Such relatively remote earthquakes are not expected to have any important impacts on one or two-storey ancient structures. This type of structure is characterized by high natural frequencies and is influenced only by high-frequency waves (resonance effect) produced by local and quasi-local events, the energy (and destruction potential) of which is rapidly attenuated away from their source.

Still, this unjustified correlation between stratigraphic discontinuities and specific earthquakes does not represent evidence of the absence of earthquake damage in an area where seismicity rates are high, and earthquakes have produced relevant changes in the geomorphology of the wider Corinthian land, as explained below. In addition, no convincing alternatives for the explanation for stratigraphic discontinuities have so far been presented, while more recent excavations tend to provide evidence of additional major destructive (seismic?) events; for example, a major destruction layer at the end of the third century (Williams II and Zervos 1985).

9.3.2 The Scenarios of Anthropogenic Effects

The scenarios of anthropogenic effects in the Corinth region cover invasions, epidemics, and religious conflicts.

In contrast to earthquakes not reported for the critical intervals, there are historical reports of invasions to Peloponnese by Heruli in 267 AD, by Ostrogoths in 395 AD, and by Slavs in the 580 s. Although various invasions had catastrophic effects in some provinces, no specific references exist to the scale of damage produced in Corinth. Perhaps there is no reason to believe that these invaders had the political and ideological stimulation, the time, and the means to turn whole towns into ruins.

A plague that hit many Roman provinces of modern Greece in 554 AD is assumed to be another reason for the decay of the Corinth region in the sixth century. Still, its impact on the town buildings is unknown (cf. Sanders 1999, pp. 574–575).

Finally, Christian zealots have been regarded as the cause of the destruction of all, but three remaining columns of the temple of Nemea (Fig. 9.1; Miller 2014). While the role of Christian zealots is recently overestimated and the Christianization of Peloponnese over-simplified (for a criticism, see Sweetman 2015), there are arguments for the role of earthquakes in the destruction of the Nemea temple. For example, if the intention of the leaders of a Christian community installed around the temple were to remove the signs of the pagan religion, none of the columns would have survived (Stiros 2020c).

9.4 An Inventory of Observable Signs of Palaeoseismic Events in the Broader Corinth Area

The most important signs of palaeoseismic events related to archaeology come from observations of coastal change. These are classified into two groups, along the North Peloponnesian coast (southern coast of the Gulf of Corinth and Saronic Gulf, cases 9.4.1–9.4.3) and along the SE coast of the Greek mainland (SE coast of the Gulf of Corinth and east coast of the Saronic Gulf (cases 9.4.4–9.4.5)).

In addition, the Nemea temple in the hinterland of Corinth is also discussed (case 9.4.6).

9.4.1 *Ancient Aigeira* (North Peloponnesus Coast)

Ancient Aigeira was built in secure area, on a 350 m high hill. The town had a harbour next to the modern site of Mavra Litharia and a rock protruding to the sea, and moreover next to which important coastal installations were built. This was an exceptional situation and a privilege for Aigeira, because soft Quaternary sediments form the south coast of the Gulf of Corinth; for this reason, Lechaion, the western harbour of Corinth discussed below, was excavated in sandy deposits.

The sea-level mark at the Aigeira harbour is 4 m above the water (Fig. 9.4), probably because of seismic events around 250 AD (Papageorgiou et al. 1993; Stiros 1998). No precise analysis of the uplift history of the harbour exists, but radiocarbon analysis of certain marine species indicates an episodic, conspicuously seismic uplift in the tenth to twelfth centuries AD. This is, however, one of the series of events that led to the present-day situation of the harbour (Papageorgiou et al. 1993).

The area's uplift history can also be derived from uplifted corals and other marine species found on two sites where solid carbonate are located on the coast, at Mavra Litharia, next to the ancient harbour, and on the nearby Trapeza

Fig. 9.4 View of the harbour of Aigeira in the late 1980s. The veranda of the house is slightly above the Roman sea-level mark. Conglomerates without a matrix originate from artificial breakwaters (see Stiros 1998)

coast (Stewart and Vita-Finzi 1996; Pirazzoli et al. 2004).

Interestingly, excavations at the Greek theatre of Aigeira provided evidence of the remodelling of the Greek theatre during Roman times. Building works seem to have been suddenly interrupted, and the theatre was abandoned. It is tempting to correlate this event with the destruction of the harbour of Aigeira, which had deprived the ancient town of its major asset (Stiros 1998).

9.4.2 *Lechaion, West Harbour of Corinth* (North Peloponnesus Coast)

Lechaion was the main harbour of Corinth in the Gulf of Corinth. Its basin was excavated into soft marine sediments (Fig. 9.5a), and for this reason, the marine basin was vested by a supporting wall built of hewn blocks. Although the harbour basin is a swamp (Fig. 9.5a), a channel (corridor) leading to the open sea is currently visible (Fig. 9.5b, c). The harbour basin area is bounded by mounds of sand from various phases of dredging of the basin (Morhange et al. 2012).

On the wall forming the channel to the sea (Fig. 9.5c), at a level of more than 1 m above mean sea level, a continuous mark of drillings by marine species, some surviving in situ, was found. AMS dating of these organisms ranged between 600 and 50 BC, indicating the period of use of the harbour. This was a surprise because due to the importance of the harbour in Roman times and to repairs in the Roman period, the harbour was traditionally assumed to have been constructed during the Roman times. Hence, AMS radiocarbon dating of seashells in harbour walls indicates that the harbour was constructed before the establishment of the Roman colony of Corinth in 44 BC. This was probably during the period when Corinth flourished in the sixth century BC. Several repairs are visible in the harbour, which has been rendered permanently unusable due to uplift and silting (Stiros 1996).

Next to the harbour, the remains of a 5th or 6th century basilica are evident (Fig. 9.5a), built in honour of Leonides who was supposed to have martyrized in this site; one of the largest early basilicas. Deformation of the foundations has been assigned to earthquake-triggered liquefaction effects (Minos-Minopoulos et al. 2015).

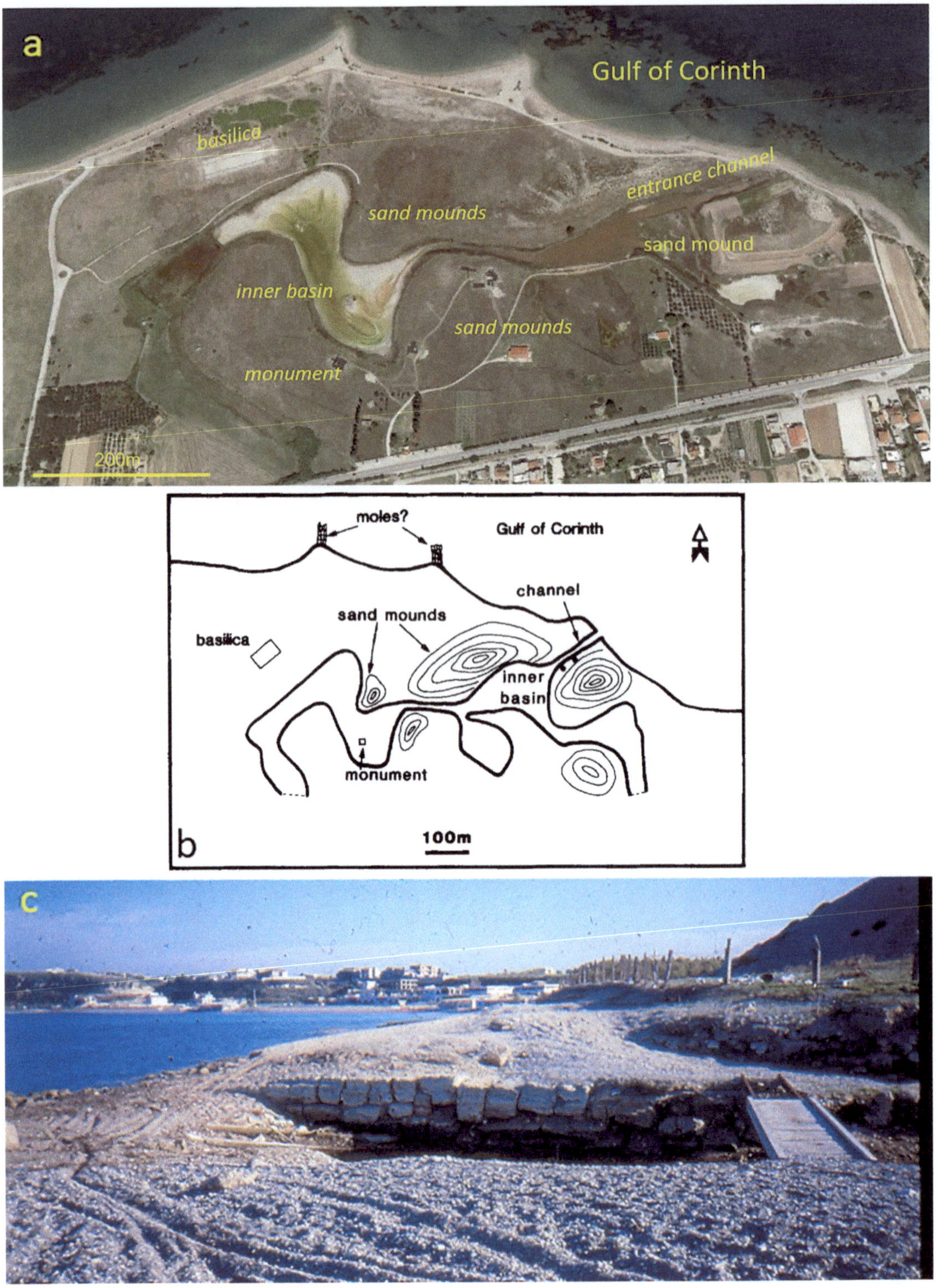

Fig. 9.5 The Lechaion ancient harbour, excavated in soft sandy formations. (**a**) Annotated Google Earth view of the harbour basin, currently a swamp. (**b**): Geometry of the harbour basin. The location of the Leonides basilica is shown to the northwest. (**c**) Eastern harbour entrance. The borders of the harbour basin were formed by walls of hewn rocks, in which a band of fossil marine organisms indicates a former sea level about 1 m higher that the present-day one, and crustal uplift. After Stiros et al. (1996). (By permission from Wiley)

Fig. 9.6 A quasi-submerged fifth-century basilica built about 1 m above the submerged floor of a building complex (hosting perhaps an Isis temple) in which about 150 glass panels were found in their casings. (After Stiros 2020a)

The palaeogeography of the harbour has been studied by Morhange et al. (2012), while recently the issue of tsunami contribution in the silting of the Lechaion harbour has been the subject of several studies with somewhat contradictory results (Kolaiti et al. 2017; Vött et al. 2018; Riddick et al. 2021).

9.4.3 *Kenchreai* (North Peloponnesus Coast)

Kenchreai was the eastern harbour of Roman Corinth, and in the middle of the second century AD, it had developed into a flourishing small harbour town, with important temples described by Pausanias (2.2.3), possibly an imitation of Ostia, the harbour for Rome (Hohlfelder 1976). Kenchreai is also known as the harbour where Apostle Paul disembarked in European land. For this reason, the ruins of a quasi-submerged church (Fig. 9.6) were usually assigned to a certain Saint mentioned in one of his Epistolas. Currently, this church is the most impressive among numerous ancient remains found quasi-submerged or submerged at two promontories enclosing the ancient harbour from north and south (see Stiros 2020a).

The ancient harbour was excavated during one of the first archaeological excavations offshore in the 1960s. The overall conclusion of the excavators is that at the time of the foundation of Roman Corinth in 44 BC, the mean sea level was about 2.3 m lower, and it has risen through three stepped land subsidence events, each with an amplitude of 0.7–0.8 m. These steps were correlated with the three earthquakes assumed to control the archaeological stratigraphy of Corinth in 77 AD, 365/375 AD, and 551 AD (see Sect. 9.3.1 and Scranton 1967; Scranton and Shaw 1978).

It was also suggested that after the first subsidence event, in the southern promontory, among other buildings now submerged, there existed a building complex, whose character is not yet clear (probably an Isis temple) and which included perhaps a yard ('fountain court'?). About 150 precious decorative glass panels in protective wooden cases were stored temporarily, standing inclined on this structure's walls and floor, along with building material and furniture. The whole complex and the promontory were submerged, and all this precious material was abandoned. This subsidence event occurred in the fourth century and was a sudden event, and an earthquake is the only reasonable explanation. Later, a church was built on top of these quasi-submerged ruins, with its floor at the height of 0.8 m above the water, but in the sixth century, a new phase of subsidence brought the floor of the church at the current sea level (Stiros 2020a).

Several modern authors have disputed the subsidence history of Kenchreai; for a review, see Stiros (2020a). The idea is that the precious panels were not suddenly drowned, destroyed, buried by debris, and forgotten, but they were abandoned because of their pagan character during a period of Christianization of the area, and the subsidence was gradual, driven by the global rise of sea levels (cf. Rothaus et al. 2008).

However, global sea-level rise since the fourth century AD was minimal, and the iconography of most glass panels had not a pagan character (Fig. 9.7), especially some of them had a Christian character. Such precious artefacts left in stacks standing on the floor along with other useful objects (furniture, tools, doors, etc.) could not have been abandoned in a coastal building complex in antiquity. The only explanation is that the whole site was covered by debris and drowned, and this can only manifest a major seismic catastrophe. There is no other reasonable explanation for this catastrophe, and only its precise dating remains unclear.

The Kenchreai panels represent the only significant example of decorative glass panels in *opus sectile* (Fig. 9.7), the origin, precise dating, and function remain unknown. They are assumed to come from Alexandria, to have been left in transit on the way to Italy, or even intended to decorate the complex in which they were found at Kenchreai (Scranton 1967).

Fig. 9.7 Two representative glass panels from Kenchreai, up to 1.5–2 m wide, after their restoration/reconstruction. Panels showing mostly marine and coastal scenes were found underwater, packed in wooden casings, probably to be used as wall decoration. After Stiros (2020a), from Ibrahim et al. (1976) and Moraitou (2017)

Fig. 9.8 Horizontal lines at a subvertical rock face at Heraion indicate four fossil shorelines, currently at the height of 3.2, 2.6, 1.7, and 1.1 m. For details, see Pirazzoli et al. (1994)

9.4.4 *Heraion (*Mainland Coast)

In the Perachora Peninsula, next to the archaeological site of Heraion, four former shorelines 1.1–3.2 m high are printed on coastal rocks (Fig. 9.8). These testify to seismic uplifts in the last 6000 years, and the last one recorded occurred during Roman times (190–440 AD).

In a lake separated by the open sea by a narrow strip of land, evidence of a prehistoric settlement was found. About 5000 years ago (Early Helladic I/II), the early phase of the settlement was abandoned, and a new settlement, about 100 m uphill, was built. There was a barrier erected with rocks along the low elevation area (Fossey 1969), probably to protect the village from a marine regression due to ground subsidence. Evidence for such temporary reversal in the sea-level trend provides some holes of limestone drilling mollusks *Lithophaga lithophaga* filled with a matrix of marine sediment. These faunal remains indicate a sequence of land uplift which led to exposed, open-hole mollusks, then land subsidence and filling of the holes, and finally, new uplift, bringing filled holes above the water (Stiros 1995a, b).

9.4.5 *Agios Vasilios* (Mainland Coast)

During the 1981 earthquakes, two lines of antithetic normal faults were formed (Jackson et al 1982; see Fig. 9.3), expressing a tendency for lowering the relief of the area between them and a tendency for expansion to the east of the gulf north of the Perachora Peninsula. In the middle of the Agios Vasilios beach, there are remains of an ancient walling system; because of the subsidence of the land by some tens of centimetres, the wall is partly submerged (Fig. 9.9; Stiros et al. 2007). This is one of the few observations of modern earthquakes' impacts on coastal archaeological sites.

9.4.6 *Nemea* (Peloponnesian Hinterland)

Apart from the Olympic Games, two other very important games were taking part in the wider region of Corinth, the Isthmian and the Nemean Games. The Nemean Games were taking place in a remote area in antiquity, where a monumental

Fig. 9.9 An ancient wall at Agios Vasilios (for location, see Fig. 9.3) submerged by a few tens of cm during the 1981 Corinth earthquakes (after Stiros et al. 2007). By permission from Elsevier

temple was built in 330 BC, probably replacing an older temple. The last Nemean Games took place in 271 BC, and the temple was found damaged by Pausanias circa 150 AD. In the fifth century AD, and possibly earlier, a Christian community was established around the temple, probably hosting a modest church. This church was soon replaced by a larger basilica, at some distance from the temple, built using some of its structural members and with a view to the remaining columns. In the last centuries, three columns of the ancient temple were left standing, while the remains of many others were found toppled down; some of them have been recently restored.

The reasons for the collapse of most of the columns of the Nemea temple (Fig. 9.1) have been a matter of debate, and two ideas have been proposed. First, the area of Nemea is deprived of strong earthquakes, and the collapse of most parts of the temple is due to deliberate destruction in the framework of the edicts of emperor Theodosius against paganism (Miller 2014). Second, a combination of seismic activity, probably manifested in the damage of the temple before the visit by Pausanias circa 150 AD, in other earthquakes, instability of the foundations imposed by various factors, including plant roots (Hill 1966), and possibly deliberate activity. A central argument for the second idea is that if Christian zealots wanted to eradicate the signs of a pagan temple, they should have destroyed all columns; it makes no sense to leave three columns standing close to the church (Stiros 2020c). In any case, the Christianization of Peloponnese was a rather complex process (Sweetman 2015).

9.5 Discussion and Conclusion

The situation with Roman Corinth and the wider region is extraordinary because

(i) This region is characterized by high seismicity rates and landscape modifications due to earthquakes, and this background has inspired and shaped archaeological investigations since the nineteenth century.

(ii) The Roman archaeological stratigraphy, characterized by destruction layers and abandonment periods, has been traditionally explained in terms of earthquake effects; yet, these earthquakes do not represent events derived from the study of ancient remains using specific criteria (for example, Stiros 1996; Hancock and Altunel 1997; Galadini et al. 2006), but they represent events known to have hit various rather remote parts of Greece (Guidoboni et al. 1994). They can be regarded as *Deus ex machina* explanations for the Corinthian archaeological stratigraphy. For this reason, most of these earthquakes are unlikely to have impacted Corinth. Hence, the Corinthian stratigraphy remains probably to some degree uncertain or biased.

(iii) The weaknesses of this approach led to rejections of earthquake scenarios, mostly in favour of theories for deliberate destruction, mostly during invasions, or because of Christianization (the cases of Kenchreai and Nemea).

Hence, a range of ideas and debates about earthquakes can be found in certain ancient areas and periods.

Apart from that, the majority of signs of earthquakes have not been obliterated with the progress of excavations and restorations because they correspond to changes in the environment of ancient structures, mostly coastal subsidence and uplift; the signs of the latter can be observed in the field, in most cases without common restrictions imposed to archaeological sites.

For these reasons, the Corinthian region represents a paradigm for Archaeoseismology, both for research and exhibition, for example, in the framework of field excursions. One may speak of the Corinthian region as an 'Archaeoseismological Park'.

Furthermore, the above discussion leads to two perspectives. A critical point for earth sciences is that earthquakes in the last two centuries do not explain the pattern of observations of coastal change, especially the transition from subsidence in Kenchreai to slight uplift at Lechaion and spectacular uplift at Aigeira. A clearer picture of the tectonics of the area is necessary. For archaeology and history, a critical point is that the impacts of such coastal changes have not yet been recognized in the historical and archaeological evolution of towns such as Corinth and Aigeira.

References

Alexandris A, Psycharis I, Protopapa E (2014) The collapse of the ancient temple of Zeus at Olympia revisited. In: 2nd Europ Conf on earthquake engin seismol, Istanbul

Avramea A (1997) Le Peloponnese du VIe au VIIIe siecle: changements et persistances. Byzantina Sorbonensia 15, Paris

Di Vita A (1995) Archaeologists and earthquakes: the case of 365 A.D. Ann Geofis XXXVII:971–976

Ellenblum R, Marco S, Agnon A, Rockwell T, Boas A (1998) Crusader castle torn apart by an earthquake at dawn, 20 May 2012. Geology 26:303–306

Evans A (1928) The Palace of Minos II. Macmillan, London

Finley JH (1932) Corinth in the middle ages. Speculum 7: 477–499

Fossey J (1969) The prehistoric settlement by Lake Vouliagmeni, Perachora. Ann Brit Sch Athens 64: 53–69

Galadini F, Hinzen KG, Stiros S (2006) Archaeoseismology: methodological issues and procedure. J Seismol 10:395–414

Gebhard E, Dickie M (2003) The view from the Isthmus, ca. 200 to 44 B.C. Corinth, 2003, vol. 20, Corinth, The Centenary: 1896–1996, pp 261–278

Guidoboni E, Comastri A, Traina G (1994) Catalogue of ancient earthquakes in the Mediterranean up to the 10th century. Istituto Nazionale di Geofisica, Rome

Hancock P, Altunel E (1997) Faulted archaeological relics at Hierapolis (Pamukkale), Turkey. J Geodyn 24:21–36

Hill B (1966) The temple of Zeus at Nemea. American School of Classical Studies, Princeton, NJ

Hohlfelder RL (1976) Kenchreai on the Saronic Gulf: aspects of its imperial history. Class J 71:217–226

Ibrahim L, Scranton R, Brill RH (1976) Kenchreai, Eastern Port of Corinth, II. The panels of opus sectile in glass. E.J. Brill, Leiden

Ivison E (1996) Burials and urbanism in late antique and early byzantine Corinth. In: Christie N, Loseby ST (eds) Towns in transition: urban evolution in late antiquity and the early middle ages, Aldershot, pp 99–125

Jackson J et al (1982) Seismicity, normal faulting, and the geomorphological development of the Gulf of Corinth (Greece): the Corinth earthquakes of February and March 1981. Earth Planet Sci Lett 57:377–397

Jusseret S, Langohr C, Sintubin M (2013) Tracking earthquake archaeological evidence in late Minoan IIIB (~1300-1200 B.C.) Crete (Greece): a proof of concept. Bull Seism Soc Am 103:3026–3043

Karakhanian AS, Trifonov VG et al (2008) Seismic deformation in the St. Simeon monasteries (Qal'at Sim'an), Northwestern Syria. Tectonophysics 453:122–147

Karcz I, Kafri U (1978) Evaluation of supposed archaeoseismic damage in Israel. J Archaeol Sci 5: 237–253

Kázmér M, Mayor B (2015) Sāfitā castle and rockfalls in the 'dead villages' of coastal Syria–an archaeoseismological study. Comptes Rendus–Geosci 347:181–190

Kolaiti E, Papadopoulos G, Morhange C, Vacchi M et al (2017) Palaeoenvironmental evolution of the ancient harbor of Lechaion (Corinth Gulf, Greece): were changes driven by human impacts and gradual coastal processes or catastrophic tsunamis? Mar Geol 392: 105–121

Korjenkov AM, Arrowsmith JR, Crosby C, Mamyrov E, Orlova LA, Povolotskaya IE, Tabaldiev K (2006) Seismogenic destruction of the Kamenka medieval

fortress, northern Issyk-Kul region, Tien Shan (Kyrgyzstan). J Seismol 10:431–444
Makris N, Psychogios T (2004) The reconstruction of the north-east corner of the Temple of Zeus at Nemea Greece, Patras, 21p, https://www.academia.edu/23952262/
Meghraoui M et al (2003) Evidence for 830 years of seismic quiescence from palaeoseismology, archaeoseismology and historical seismicity along the Dead Sea fault in Syria. Earth Planet Sci Lett 210:35–52
Miller S (2014) Theodosios II and the Temple of Nemean Zeus. http://www.arxeion-politismou.gr/2018/03/Theodosios-Nemea-Miller.html
Minos-Minopoulos D, Pavlopoulos K, Apostolopoulos G, Lekkas E, Dominey-Howes D (2015) Liquefaction features at an archaeological site: investigations of past earthquake events at the early Christian Basilica, Ancient Lechaion Harbour, Corinth, Greece. Tectonophysics 658:74–90
Moraitou G (2017) Coloured, transparent figures from the harbor of Kenchraiai (in Greek). Archaiologia 123:28–37. https://www.archaiologia.gr/wp-content/uploads/2018/03/123_28-37.pdf
Morhange C, Pirazzoli P, Evelpidou N, Marriner N (2012) Late holocene tectonic uplift and the silting up of lechaion, the western harbor of ancient Corinth, Greece. Geoarchaeology 27:278–283
Nikonov A (1998) Strike-slip deformations in the Amu Darya delta region: archaeological evidence. Doklady Earth Sci 361A:804–808
Nur A, Ron H (1997) Armageddon's earthquakes. Int Geol Rev 39(6):532–541
Papageorgiou S, Arnold M, Laborel J, Stiros S (1993) Seismic uplift of the harbour of ancient Aigeira (Central Greece). Intern J Naut Archaeol 22:275–281
Papazachos B, Papazachou C (1997) The earthquakes of Greece. Zitis, Thessaloniki, p 304
Pirazzoli P et al (1994) Episodic uplift deduced from Holocene shorelines in the Perachora Peninsula (Corinth area, Greece). Tectonophysics 229:201–209
Pirazzoli P, Stiros S, Fontugne M, Arnold M (2004) Holocene and quaternary uplift in the central part of the southern coast of the Corinth Gulf (Greece). Mar Geol 212:35–44
Rapp G Jr (1987) Assessing archaeological evidence for seismic catastrophes. Geoarchaeology 1:365–379
Riddick N, Boyce J, Reinhardt EG, Rothaus R, Chomicki KM, McCarthy F (2021) Multi-proxy palaeoenvironmental record of coastal tectonic uplift and abandonment (ca. 6th c. CE) of Lechaion's inner harbour, ancient Corinth, Greece. Quat Sci Rev 267, no. 107080
Rodríguez-Pascua M, Pérez-López R, Giner-Robles J, Silva P, Garduño-Monroy V, Reicherter K (2011) A comprehensive classification of earthquake archaeological effects (EAE) in archaeoseismology: application to ancient remains of Roman and Mesoamerican cultures. Quat Int 242:20–30
Rothaus R, Reinhardt E, Noller J (2008) Earthquakes and subsidence at Kenchreai: using recent earthquakes to reconsider the archaeological and literary evidence. In: Caraher WR, Hall LJ, Moore RS (eds) Archaeology and history in medieval and post-medieval Greece: studies on method and meaning in Honor of Timothy E. Gregory. Ashgate, pp 53–66
Sanders G (1999) A late Roman Bath at Corinth: excavations in the Panayia field, 1995-1996. Hesperia 68:441–480
Scranton R (1967) Glass pictures from the sea. Archaeology 20:163–173
Scranton R, Shaw J (1978) Changes in relative sea-level. In: Scranton R, Shaw J, Ibrahim L (eds) Kenchreai, the eastern port of Corinth I, Appendix E. Leiden, Brill, pp 144–147
Stewart IS, Vita-Finzi C (1996) Coastal uplift on active normal faults: the Eliki Fault, Greece. Geophys Res Lett 23:1853–1856
Stiros S (1995a) Palaeogeographic reconstruction of the Heraion-Vouliagmeni Lake coast since the early Helladic times. Annu Br Sch at Athens 95:17–21. & plate 4
Stiros S (1995b) Palaeogeographic reconstruction of the Heraion–Vouliagmeni Lake coast since Early Helladic times. Annu Br Sch Athens 90:17–21
Stiros S (1996) Identification of earthquakes from archaeological data: methodology, criteria and limitations. In: Stiros S, Jones R (eds) Archaeoseismology, Fitch laboratory occasional paper 7, Oxford, pp 129–152
Stiros S (1998) Archaeological evidence for unusually rapid Holocene uplift rates in an active normal faulting terrain: Roman harbour of Aigeira, Gulf of Corinth, Greece. Geoarchaeology 13:731–741
Stiros S (2020a) Coastal subsidence, destruction layer, and earthquakes from an underwater archaeological excavation: Kenchreai, the eastern harbor of Roman Corinth, Greece. Mediterr Geosci Rev 2: 283–297
Stiros S (2020b) Was Alexandria (Egypt) destroyed in AD365? A famous historical tsunami revisited. Seismol Res Lett 91:2662–2673
Stiros S (2020c) Monumental articulated ancient Greek and Roman columns, temples, and earthquakes using archaeological, historical, and engineering approaches. J Seismol 24:853–881
Stiros S, Pirazzoli P, Rothaus R, Papageorgiou S, Laborel J, Arnold M (1996) On the date of construction of Lechaion, the western harbor of Corinth, Greece. Geoarchaeology 11:251–263
Stiros S, Psimoulis P, Pitharouli S (2007) Geodetic constraints to the kinematics of the Kapareli fault, reactivated during the 1981 Gulf of Corinth earthquakes. Tectonophysics 440:105–119
Sweetman R (2015) The Christianization of the Peloponnese: the case for strategic change. Ann Brit Sch Athens 110:285–319

Vött A, Hadler H, Koster B, Matthes-Schmidt M et al (2018) Returning to the facts: response to the refusal of tsunami traces in the ancient harbour of Lechaion (Gulf of Corinth, Greece) by 'non-catastrophists'–reaffirmed evidence of harbour destruction by historical earthquakes and tsunamis in AD 69–79 and the 6th cent. AD and a preceding pre-historical event in the early 8th cent. BC Zeitschr Geomorph 61:275–302

Williams CK II, Zervos OH (1985) Corinth, 1984: east of the theater. Hesperia 54:55–96

10 Research on Historical Seismicity in Vienne (France): From Archaeoseismological Observations to the Archaeological Identification of 'Viennese' Earthquakes in Antiquity and the Middle Ages

Bruno Helly, Alain Rideaud, and Benoît Helly

Abstract

In the Roman city of Vienne (France), the archaeoseismological observations revealed a series of disorders, repairs, and reconstructions characteristic of ground movements: at the temple of Augustus and Livia, in the Roman Forum. These observations have led archaeologists to focus more on a rich array of archaeological evidence. This includes the interpretation of the two states of construction of the temple of Augustus and Livia, as well as those that we have on the forum: – the temporary stopping of the construction of a large horrea located on the edge of the Rhône; – the renovation of the urban plan and the development of the construction of large urban residences. These observations made it possible to formulate the hypothesis that it was likely an earthquake that had caused the almost total destruction of the city, leading to the reconstruction and transformation of most monuments and houses. We have sought to extend these observations to buildings from the mediaeval period. Thus, we have observed characteristic disorders and reconstructions on the church and cloister of Saint André le Bas, at the cathedral church of Saint-Maurice, and finally at the Château de la Bâtie, built by the counts of Albon in the middle of the twelfth century.

We could compare our hypotheses with the findings formulated by geologists on the existence of a possible zone of tectonic disturbances in the lower valley of the river La Gère, possibly originating from an 'orphan' fault. We believe that the activation of this rift would have taken place at the beginning of the Roman imperial era, possibly in the sixth century AD, if one believes a testimony of Sidoine Apollinaire, and again in the course of the fourteenth century? To verify these hypotheses, once again, archaeologists and archaeoseismologists will try to converge their respective research.

Keywords

Vienne (France) · Civitas Viennensis · Historical seismicity · Reconstructions after an earthquake

B. Helly (✉)
Directeur de recherche au CNRS (émérite), Maison de l'Orient, Lyon, France
e-mail: bruno.helly@mom.fr

A. Rideaud
Architecte DPLG, Tournus, France

B. Helly
Ingénieur d'études Service Régional de l'Archéologie, DRAC Auvergne-Rhône Alpes, Lyon, France

The city of Vienne, located 35 km south of Lyon, was a major town in antiquity. It was established on the east and west bank of the Rhône River,

L. Pecchioli (ed.), *Archaeoseismology*, Natural Science in Archaeology,
https://doi.org/10.1007/978-3-031-28303-1_10

including the actual municipalities of Vienne, Saint-Romain-en-Gal, and Sainte-Colombe. More than a hundred years of archaeological research have revealed the city's importance, and the archaeological activities continue under the control of the Direction régionale de l'archéologie from the Ministère de la Culture.

Vienne was founded probably in the fifth century BC by the Gallic people the Allobroge's as their capital and was attached to the Roman province of Transalpine in the last years of the second century BC and received colony status under Caesar (Helly 2013, 116–148). Despite the continuity of the urban occupation, modern Vienne and the two cities on the other bank of the Rhône have remarkably preserved remains of monuments from Roman times. The archaeoseismological observations on these monuments and other buildings revealed a series of disorders, repairs, and reconstruction consequences of ground motions, the so-called Earthquakes Archaeological Effects (EAEs), that can be attributed to seismic events (Fig. 10.1).

10.1 Observations of Earthquakes Archaeological Effects (EAEs)

The temple of Augustus and Livia, following the plan of the Roman temple, a rectangular pseudo-peripteral building of the Corinthian order (raised on a high podium, with a portico in front), is already well studied by archaeologists who differentiated two stages of construction (Gros 1996, 158–159). The first stage, dated from the Augustan period (around 20 BC), retains traces of the collapse of the east part of the building, the portico. We can observe an offset in column alignment, rotations of bases, cracking of architraves, and disorders which are east-west oriented. The design of the cracks follows a parabolic curve upward facing, which creates strong suspicion of ground motion of seismic origin (Helly and Rideaud 2016). The western part of the monument, occupied by the *cella,* shows a twisting of the west wall, and the bases of the columns, which remain standing, show rotations and breaks in alignments.

The temple was repaired, and the east part with the portico and the fronton were rebuilt with new material (Fig. 10.2). In particular, the dedicatory inscription on the pediment was changed with the addition of the name of Livia, the wife of Augustus, which was deified by the emperor Claudius in 42 AD.

Similarly, traces of EAEs were identified on the two remaining arches of the Roman forum: block displacements, dipping broken corners or chipping marks, rotations, spills and gap from the bottom to the top of the pillar are visible on the first state arch, dating from the end of First (?) Century BC (Fig. 10.3). All these disorders testify to vertical movement. Characteristic is the break of continuity with the neighbouring arch whose vault was rebuilt with reused materials and special techniques of stereotomy, with blocks in crossettes (Adam 1984, 184–187), which reflects the aim to achieve better resistance (Fig. 10.4).

Initially observed on the monuments still in place, EAEs were also identified in the documentation of the old excavations. These signs of the disaster were visible in the destruction-reconstruction of private buildings: rue des Colonnes, place St Pierre, rue de Bourgogne, where the shop of a pottery merchant collapsed and was destroyed by fire at the end of the reign of Tiberius. In the forum sector, the so-called "Cybele garden" exists as a ruined Tiberian habitat whose remains were found under the assembly hall built at the time of Emperor Claudius (Helly 2013, 116–148).

In the south part of the city, on the left bank of the Rhône, an interruption was observed in the construction of the large Augustinian warehouses (*horrea*) following a transgression of the river, a tsunami-like wave (Rodríguez-Pascua et al. 2011, 27: 'Roof and vault collapses of inhabited houses often start fires by the burning of the wood fabric of most pre-modern roofs. Flash flood events triggered by severe earthquake damage of ancient earth dams can also be a source of information, although the interpretation is complicated'. On the right bank of the Rhône, in the artisanal and residential district of the city, we have observed folded and fractured pavements. On the site of the 'Bourg' in Sainte-Colombe, the north-south walls

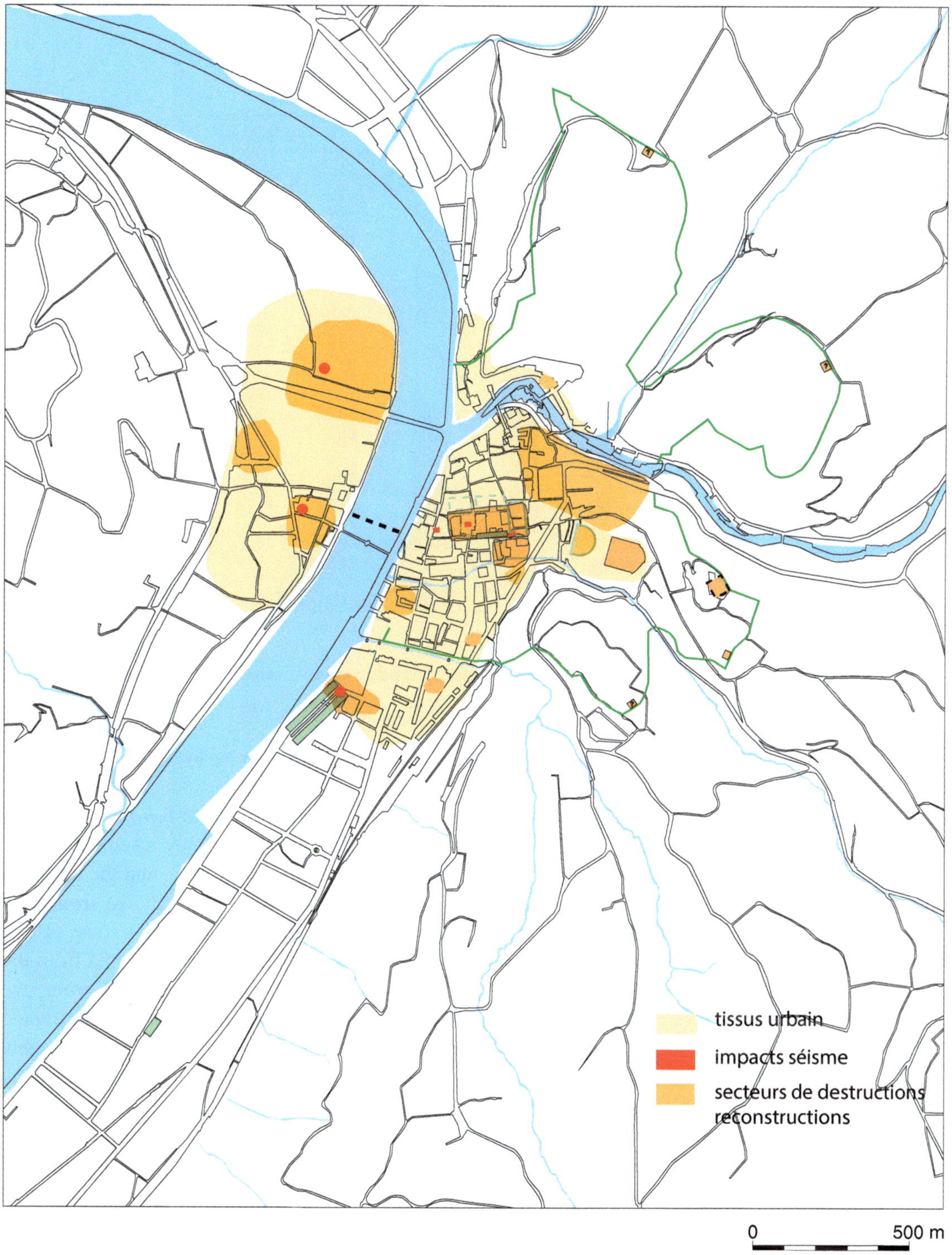

Fig. 10.1 General plan of the Roman Vienne with the location of observation points of disorders, repairs, and reconstructions; vignette: geographic location of Vienne. (Map by Claire Marcellin and Benoit Helly)

Fig. 10.2 The south elevation of the temple, with the two states of the monument (Drawing by Benoit Helly)

of a warehouse built under Augustus were destroyed at the end of the reign of Tiberius (Fig. 10.5). The archaeologist noted penetrative fractures in masonry blocks. The repair of walls takes place under Caligula and a total reconstruction under Emperor Claudius (Clément 2018 and Fig. 10.6 and Clément 2020).

10.2 Secondary Effects: Repairs and Reuse of Materials

The reconstruction, repair, and reinforcement of earthquake-related damaged structures can be interpreted as secondary evidence for earthquakes (Rodríguez-Pascua et al. 2011). The rebuilding of the temple of Augustus and Livia is characterized by a change in the building material: we observe in the restoration of the temple, for the most part, the use of a limestone Urgonian called the stone of Seyssel (Haute-Savoie), which is different from that previously used (a lighter stone from the south of France), and the insertion of a harder limestone called 'choin' from new quarries situated north of Lyon. The use of this new material was generalized in Vienna from the reign of Emperor Claudius. At the Forum, we noted that the vault of the arch was reconstructed using another technique: the blocks of the new arch are cut outside in crossettes, multiplying the points of fixations between the blocks of the arch. The care given to the reconstruction of the elements of the collapsed arch seems to correspond to the aim to strengthen the resistance of the structure to shocks.

We have the indirect testimonials of the rehabilitation of the aqueducts and the water distribution: eight identical inscriptions recounting the donation by a Sulpicia Censillia of 50,000 sesterces 'for the perpetual preservation of these

Fig. 10.3 The remaining arches of the forum entrance: right, the remaining arch from the first century BC; left, the arch with reconstructed vault (Photo, Alain Rideaud)

signs' (*titulos/tuandos*), a dedication commemorating the generosity of his father D. Sulpicius Censor and Q. Gellius Capella, two *quattuorviri* having offered new waters supplies for the inhabitants of Vienne. On this occasion, repair of the fountain in the forum must have occurred and features his inscription (Helly 2017, 92).

It is likely that for a while, it was necessary to live in ruins. As a result of the disaster, the cleaning of the ruins and the recovery of materials were not systematic and immediate. Some buildings were simply levelled off, as shown by a Th. Delorme excavation in the nineteenth century on the site called 'Cybele' in Vienne, a house which burned, yet everything remained in place (Helly 2013, 148).

The architectural elements are obviously recovered and reused, both in public monuments and in private residences. The archaeologist A. Desbat even emphasizes in relation to the Pierres-Dorées house that the recovery of materials has been systematic, from the tiles of the roof to the tesserae of the mosaics (Desbat et al. 1994, 201–205).

There have been many temporary repairs and 'crafts' to maintain some buildings and houses. In

Fig. 10.4 Drawing of the reconstructed arch, south face (TPLM 3D, Chasse-sur-Rhône)

Saint-Romain-en Gal, for example, the mud-brick roofs of an Augustean house.

The Painting House is raised mortar bondage, different from the original wall (Faure-Brac 2006, 444). Temporary repairs, for temporary occupation of 10 years, the house was levelled and backfilled in the middle of the first century to make room for the Palace of the Mirror (Fig. 10.6). Still, in Saint-Romain-en-Gal, the repair of the water supply of the Pierres-Dorées house dates from the same period. This building will give way after the collapse to the house at Grand Péristyle (40–60 AD).

Similarly, we can note the recovery of materials, resumption of elevations of the house with paintings, for example, at Sainte-Colombe (rue Garon, ZAC du centre), the negative traces of the stone jambs of a door of a Tiberian house that was then backfilled (Faure-Brac 2006, 323–325). Elsewhere, the sorting of materials was done: thus, painted coatings and earth elements of half-timbered or cob houses, most of the time burned, were set apart to constitute embankments. The analyses of recycled and reused architectural elements at different points of the city show such embankments made up of earthen architectural elements and tiles of roofs (Frascone 2013). The important previous stratigraphy located several metres deep on the substratum on which the constructions of the most important monuments had to be supported required special arrangements. Thus, several temples are then placed on thick rafts resting on caisson foundations connected sometimes by lowered vaults (most often with lost wood), whose walls are deeply anchored in the natural ground while crossing the previous levels (Fig. 10.7).

10.3 The Reconstruction of the City: A Reasoned Town Planning Programme

These observations have led archaeologists to give more attention to the rich array of archaeological evidence of destructions and

Fig. 10.5 Map of the constructions at Sainte-Colombe: destructed walls in red (Map by Benjamin Clément, Archeodunum)

reconstructions. That includes the interpretation of the two states of construction of the Temple of Augustus and Livia, as well as those that we have on the forum; the temporary stop of the construction of large *horrea* located along the Rhône river; the renovation of the urban plan and the development of the construction of large houses. Archaeologists have also further exploited indirect sources such as epigraphy and texts. The inscriptions relating to the repair of the aqueducts, the repair of fountains, and a clock were financed by wealthy citizens (Helly 2013, 114–148).

We noted above the reconstructions that have been carried out on the temple, on the forum, and

Fig. 10.6 Reconstruction of elevations of the house with paintings at Saint-Romain en Gal (Photo, Benoit Helly)

other buildings in the city. Such operations are also found on other monuments, like the fountain in the forum, with his inscription: 'Decimus, of the Voltinia tribe, edile, quattuorvir', has offered new water sources and conduits of these waters through their estates to the settlers of Vienne. For the perpetual preservation of these signs, Sulpicia Censilla, daughter of Decimus, has ordained by will the award of a sum of fifty thousand sesterces, and even the transformation of the theatre, on which we recognize two successive phases, first one Augustan and secondly one Claudian, in parallel with the rebuilding of the temple. We also noted the adaptation of construction techniques and the change of stone material, which resulted in the opening of new quarries of a limestone portlandien of South of the Mounts Jura, called 'choin', on the northern part of the territory of the city to meet the needs of freestone and lime, on the right bank of the Rhone, especially the north of department Isère and the south of Ain (Helly 2013, 114–148).

All this reveals ambitious urban policy based on the research of all means necessary for the city's reconstruction. The city officials have undoubtedly been mobilized to help this reconstruction. Thus, many inscriptions from the second half of the first century AD mention the acts of benefactors directly related to the repairs (restoration of fountains and aqueducts, restoration of a clock), but also of embellishment, donation of bronze tiles of a temple, statues, porticoes, and so on. It is also necessary to suppose other probable donations, for example, the reconstruction of the theatre, and financial aid of Antonia Minor, of which a statue was enthroned in the new edifice built under Claudius. Moreover, it cannot be ruled out that a supply of workforce in the form of one *vexillatio* has been granted to Vienne. The evidence of the presence of barracks and *canabae* built under Claudius or Nero south of Sainte-Colombe would go in this direction (Helly 2018).

A reconstruction of this size requires very important funding. The main element that favoured the city's reconstruction was obtaining the Italic right as financial compensation. A simple temporary suspension of taxes (often for 5 years) would not explain the particularism of

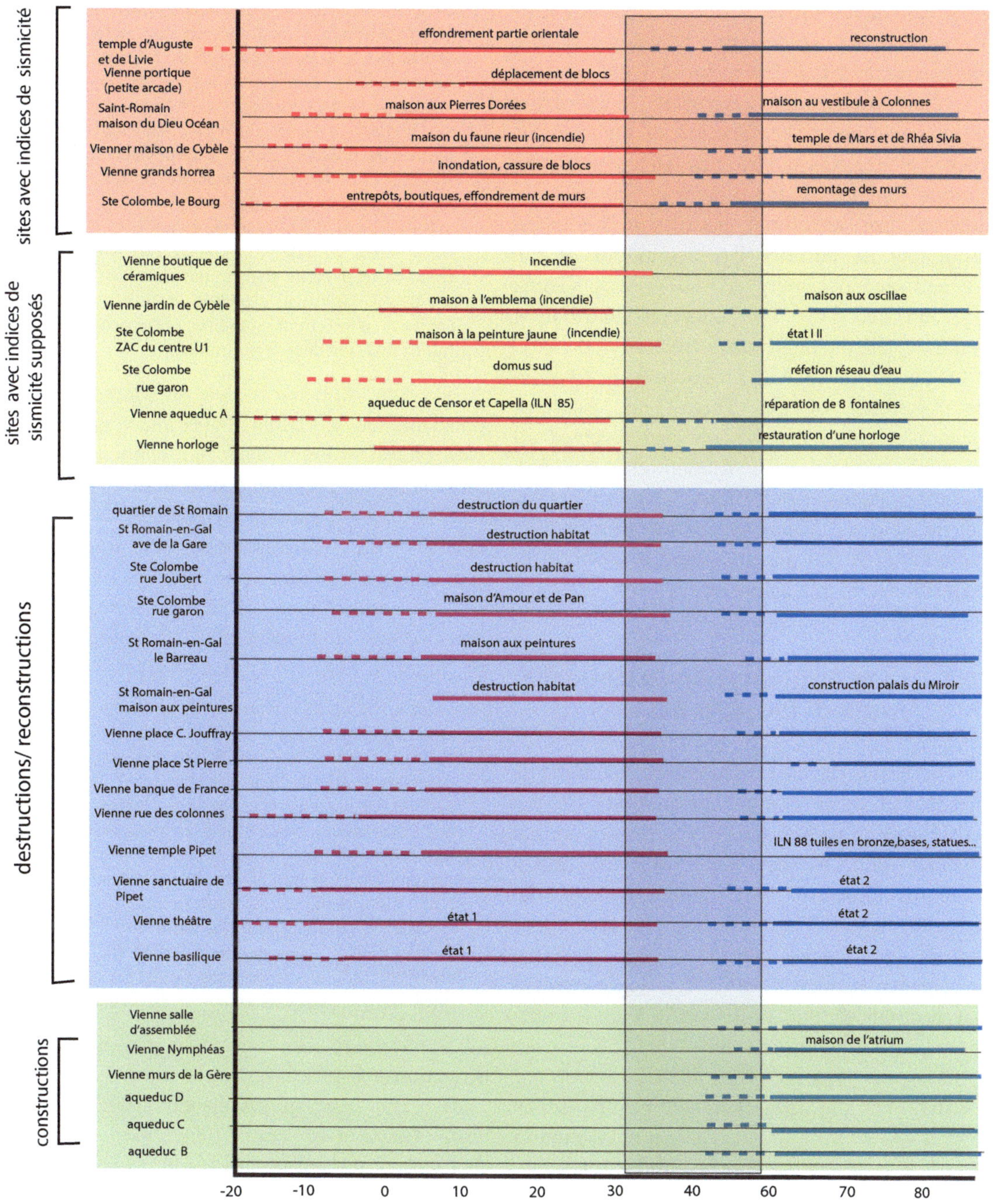

Fig. 10.7 The chronological range of all the destructions and reconstructions in Vienne's Roman city: end of Emperor Caligula's reign–beginning of Emperor Claudius's reign (compiled by Benoît Helly)

the status of Vienne and his new political organization attested in the first and second centuries AD. The *Ius italicum* was likely granted by Caligula to the city, with the tax relief it allowed, with full exemption to pay tax to Rome.

Obtaining *Ius italicum* under Caligula had great political consequences. We notice a complete change in the political organization of Vienne at this date: the passage of the *quattuorivrat* to the *duumvirat* and the creation of the *triumvirat* (*locorum publicorum persequendorum*). The function of the *triumvir* was to collect taxes and tributes. This new administrative function of the Viennese city was not punctual, nor occasional, because it is attested throughout its territory by inscriptions of the first and second century AD (*ILN*, V 3, Vienna 317, 786, and 848). This financial contribution would explain the rapid reconstruction of the city. In his speech of 48 AD, pronounced in Lyon in the Assembly of the Three-Gaul, Emperor Claudius declares that Vienne had received all the rights granted to a Roman city (*solidum ciuitatis romanae beneficium*). Claudius praises the richness of *Viennese architecture*, which implies that it has already been at least partly reconstructed at this date.

10.4 Late Antiquity: An Earthquake Around 450–470 AD

Our observations also focused on monuments after Roman times. For the late antiquity and the early Christian period, the only witnesses are the churches and monasteries. Still, the many transformations they have undergone over the centuries currently preclude a direct reading of damage or repairs attributable to tectonic movements. However, we have a written source that mentions earthquake effects in Vienne in the second half of the fifth century AD.

The poet Sidoine Apollinaire, bishop of Clermont-Ferrand from 470 until 476, writes around 472 to St. Mamert, Bishop of Vienna in Dauphiné, to announce that he comes, according to his example, to institute the Rogations in his diocese. He reminds him that the city of Vienne (*civitas tibi caelitus credita*) was desolated by prodigies of every kind (*Lettres*, VII, 1, col. 102):

> At that time, Mamert, bishop of Vienna, removed from the city of Vienne by his tears and his prayers, a scourge which threatened it. Because frequent fires, incessant earthquakes, and nocturnal rumblings threatened some prodigious and fatal event for the ruin of the whole city, and while the meetings of men were numerous, one could see the appearance inside the city of wild beasts of the forest: wolves, bears, and deer, naturally subject to fear, penetrated through the narrow doors to the vast expanse of the forum. What? While the solemn feast of the impending Holy Saturday celebrates by which it is customary to celebrate the feast of the Resurrection of our Lord, a public building, which, at the very top of the Viennese city, the height of its ridge raised infinitely, began in the twilight to embrace frightening flames; the manifestations of joy due to [spiritual] elevation are interrupted by the news of the critical situation; as the people were full of fear, the church is emptied.

St. Avit, who succeeded St. Mamert on the episcopal see of Vienne in 490, and was therefore well placed to know what had happened in that city during the preceding generation, recounts the institution of rogations by Mamert as recourse against the misfortunes of time ('De rogationibus', *Patrologie latine* of Migne, t. LIX, ed. Peiper, 1883, p. 108 and 'De consolatoris laude castitatis', *Patrologie latine* of Migne, t. LIX, ed. Peiper, 1883, p. 373):

> I know, 'he says', that many of our fellow citizens remember the terrors of that time [when the Rogations were instituted in the year 452 of the Christian era)]. In fact, frequent fires, continual earthquakes, and nocturnal sounds seemed to presage a stupendous pyre of the expiring universe. For the beasts of the forest, become servants, and presented themselves in the many assemblies of men. Who, then, seeing frequent fires (*in crebris ignibus*), would not fear the rains [of flame] of Sodom? Who, then, in the midst of the shaken elements, would not see hanging on his head the threat of roofs ready to collapse or lands crumbled?

St. Mamert then conceives the thought of Rogations. A procession celebrated 3 days preceding the feast of Ascension. Two centuries later, Gregory of Tours (*Hist. Francorum*, II, 34) repeats this information with the mention of 468 AD (Lucas 2014, 60–90).

These texts were the subject of a critical study by the specialist of historical seismicity at the Bureau de Recherches Géologiques et Minières (BRGM), Jérôme Lambert, in a note for the French Commissariat for Atomic Energy (Department of Nuclear Safety) in January 1985. J. Lambert concludes this study:

> All the sources consulted allow to establish that only the texts of Sidoine Apollinaire and Saint Avit have a value of serious proof. Although the exact date cannot be specified, their analysis makes it possible to determine the existence of a series of tremors felt in Vienna, poorly individualized, of a maximum possible intensity of V to VI MSK, and unknown extension. In addition, the hypothesis of a close epicenter could be accepted. (Lambert 1985).

The seismic events attested by the testimony of Saint Mamert before 470 may have been the cause of the destruction of the first church and *martyrium* of Saint Ferréol, a funerary basilica built in the fourth century on the place of the martyrdom of Saint Ferréol, directly on the shore of the Rhone on the right bank of the river (church of Saint-Romain en Gal). This *martyrium* of St Ferréol was replaced by a second church, built by the bishop of Vienne, Saint Mamert at some distance from the first, according to the narration of Gregory of Tours:

> 'And after reading (the inscription), I asked the porter why this inscription, he replied, The Basilica of St. Martyr Ferréol was once elevated directly at the edge of the Rhone. Subsequently, as the portico on this side began to collapse due to the violence of the river flow, a wise. A Bishop named Mamert, who was then running the church in Vienna, planning the next ruin of the building, had built another basilica, a monument elegant and of the same size, with a pious intention because he wanted to transfer the sacred body of the martyr'.

This destruction of the first church of Saint Ferréol may be contemporary to the reported destructions by the earthquake which occurred under the episcopate of Saint Mamert, and makes plausible the possibility that a seismic shock caused a wave in the bottom of the river, as already mentioned (Reynaud et al. 2019).

10.5 An Earthquake in the Middle Ages Between 1250 and 1350 AD?

Other observations of disorders we can hypothetically interpret as characteristic EAEs carry us back to the Middle Ages. In the Romanesque church of Saint-André-le-Bas, built in the 11th AD as a church of a Benedictine monastery and rebuilt in 1152 in the later Romanesque style, we have noted rotations of the bases of Romanesque pillars, one of them which is inserted the first stone of the (re)?-construction of the church mentions the architect, Guillaume, son of Martin. The cloister linked to this church, apparently Romanesque in appearance (Formigé 1947; Barruol 1992), was, in fact, rebuilt and enhanced with the reuse of previous material from several sites (Boissard 2011). The arrangement of the colonnade and its capitals is, to say the least, composite with capitals with a diameter greater than that of the baluster, for example, in Fig. 10.8. The direct relationship of this reconstruction was not established. Still, it should be noted the chronological concordance with seismic phenomena observed in the city at this time (Formigé 1947; Barruol 1992).

In the Gothic Cathedral of St Maurice, on a basilica plan with three aisles and an apse, but no ambulatory or transepts, built in 1052 AD, in which the Council of Vienne convened in October 1311 by Pope Clement V abolished the order of the Knights of the Temple, is observed a typical deformation of the three south-west spans at the basis of a pillar and at the vaults (Fig. 10.9). That may suggest the collapse of the facade and then went on to build three new west spans and the west front after the late thirteenth, early fourteenth century AD (Cavard 1978, Bouticourd 2012, vol. me 1, 67–82). In the same Cathedral Saint-Maurice, an inscription dated between 1292 and 1295 AD relating to the foundation of Guigues, chanoine of the cathedral, mentions a donation of money to repair the vaults of the church (Favraud et al. 1990, 40–43).

Other possible EAEs are visible on the ruins of Château de la Bâtie, built in 1225 AD, particularly a fracture oriented South-West/North-East, which seems to cross the entire construction, from its visible foundations to the top of the wall (Fig. 10.10). This large so-called sword stroke is straight in plan and elevation through the entire main body of the castle and seems to have also affected the rock underlying the construction. One could think that a landslide could cause such a disorder, but the crack is more like a parabolic line, whereas this 'sword stroke' is

Fig. 10.8 The reconstructed cloister of the abbey of St André le Bas (Photo, Benoit Helly)

clearly vertical. An earthquake may well be at the origin of the fracturing of this construction, as we have often noted. If an earthquake is the cause of such a disorder, other observations built at the same time should confirm this, but it can also be several successive events, ending up damaging the building for lack of maintenance, for example, this hill, by 'site effect', could amplify shaking. We know that mediaeval written sources say little about earthquakes (Demotz 1973, 17–28 and Demotz 1975, 55–60).

Therefore, as we were able to achieve for the destruction reconstructions of the first century AD, we present a table with the chronology of the destruction and reconstruction in mediaeval times (Fig. 10.11).

10.5.1 New Research Perspectives for the Archaeological and Seismological Studies

The description of the disorders that we can interpret as EAEs and the secondary and indirect effects, such as the reuse of materials, repairs, and reconstructions that affected monuments

Fig. 10.9 South nave of the Cathedral Saint-Maurice (Lidar survey by Marc Jadeau, Méga 3 DIM, Vienne)

Fig. 10.10 The castle of La Bâtie (Photo, Alain Rideaud)

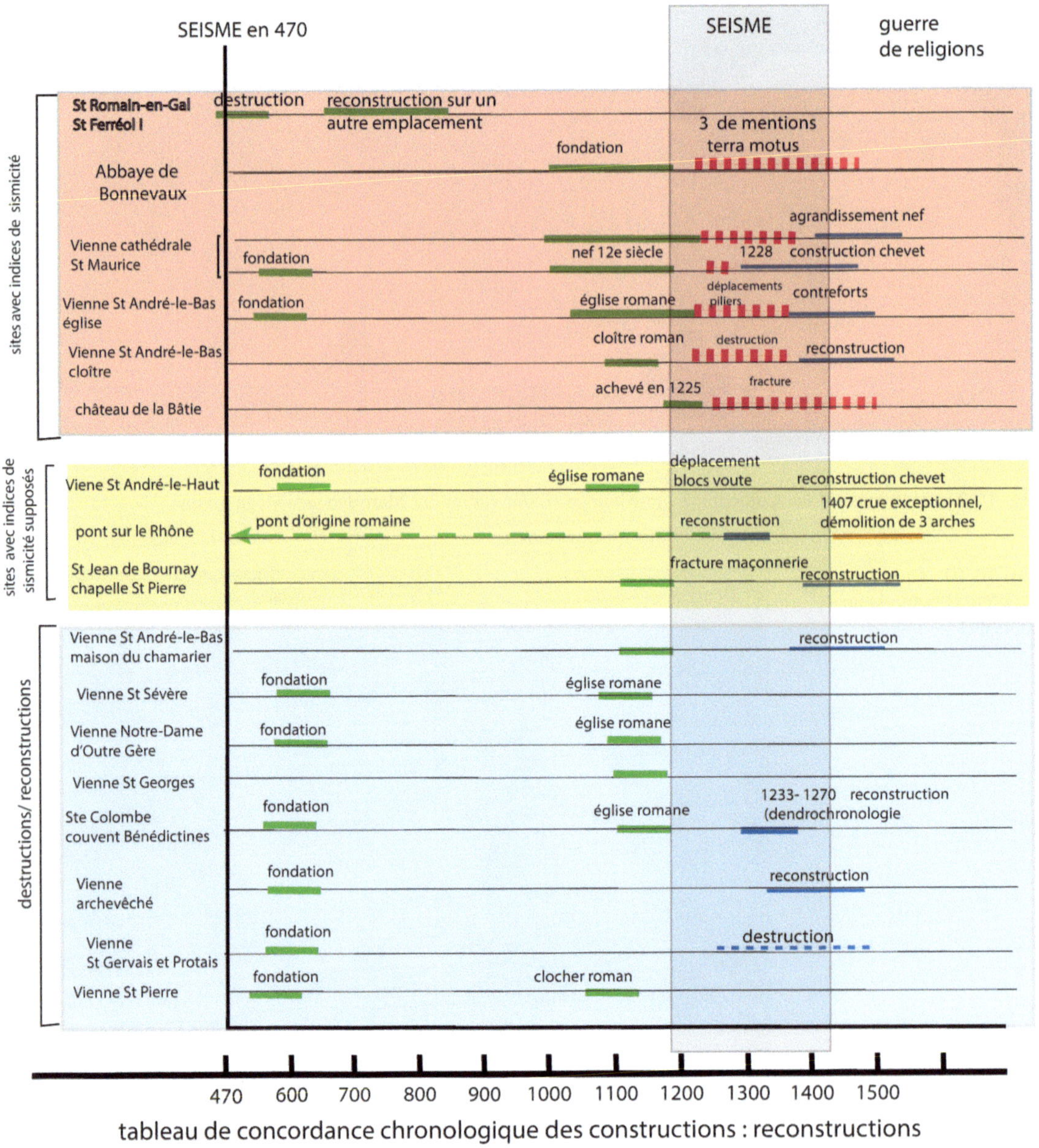

Fig. 10.11 Chronology of the destructions and reconstructions in mediaeval times (compiled by Benoît Helly)

and other buildings in Vienne in three different periods, is just the beginning of research which is to be developed in the future. First, knowledge of the characteristics EAEs is likely to apply in future excavation operations. Secondly, it is necessary to undertake more precise studies on architectural structures and soils. Further, we need geo-seismic studies of possible faults on the river la Gère and also supplementary research on the geological context. This constitutes the goals of a scientific programme entitled 'Archéosismicité et Environnement à Vienne'. Multidisciplinary, it includes archaeologists, archaeoseismologists, geophysicists, geologists, and geochemists. The programme has recently commenced under the responsibility of Madame Laurence Audin (ISTERR, University of Alpes-

Grenoble) with the financial support of the Region Auvergne-Rhône-Alpes.

Acknowledgments We thank all the colleagues with whom we had enlightening discussions on the history of the monuments of Vienne. We are indebted to Marc Jadeau, Méga 3 DIM, Vienne, for the laser pictures of the vaults in the Cathedral Saint-Maurice, and Benjamin Clément, Archéodunum, for the observations he made in the excavations of the site of the Bourg' in Sainte-Colombe.

References

Adam JP (1984) La construction romaine, matériaux et technique, éditions Picard

Barruol G (1992) Vienne, Saint-André-le-Bas, in Dauphiné roman, éd. Zodiaque, pp 113–156

Boissard E (2011) Vienne (Isère), Cloître de Saint-André-le-Bas, rapport de fouille, Archeodunum, Chaponnay. 123

Bouticourd E (2012) Vienne (Isère), Ancienne cathédrale Saint-Maurice de Vienne, Façade Nord, Travées 5 à 13, rapport de fouille (étude de bâti), Archéodunum, Chaponnay. 3

Cavard P (1978) La cathédrale Saint Maurice de Vienne. Blanchard Frères, Vienne, p 188

Clément B (2018) Sainte-Colombe, une mini-Pompéi des bords du Rhône, Archéologia N°564, avril 2018, p 34–43. Dijon

Clément B (2020) Un quartier antique de la colonie romaine de Vienne figé par les flammes. Premiers résultats des fouilles du Bourg à Sainte Colombe, in Bulletin de la SociéInsérer une formule tté d'archéologie classique, Revue Archéologique (XLX, 2018–2019), 141–191

Demotz B (1973) Un château méconnu: la Bâtie de Vienne: Bulletin de la société des Amisde Vienne, N° 69 fascicule N°3, Vienne, 1974

Demotz B (1975) La Bâtie de Vienne, une résidence épiscopale devenue une puissante forteresse, Archéologia N° 88. Dijon 1975:55–60

Desbat A, et al (1994) Les fouilles de la maison des Dieux Océans à Saint-Romain-en-Gal, 55e supplément à Gallia, CNRS édition, Paris. p 276

Faure-Brac O (2006) Carte Archéologique de la Gaule, le Rhône, pré-inventaire archéologique, 69/1, Maisons des sciences de l'Homme, Paris, 2006: Sainte-Colombe p. 307–357, Saint-Romain-en-Gal, p. 405–506

Favraud R, Michaud J, Mora, B (1990) Corpus des inscriptions de la France médiévale, 15.La ville de Vienne en Dauphiné, CNRS Éditions, 40–43, n° 38

Formigé J (1947) Le cloître de Saint-André-le-Bas. In: Pierre W et al (eds) Le cloîtrede Saint-André-le-Bas à Vienne, inscriptions chrétiennes du musée d'art chrétien, pp 37–42

Frascone D (2013) Vienne, 2 cours Brillier, rapport de diagnostic, INRAP, Bron, 2013. 120 p

Gros P (1996) L'architecture Romaine du début du IIIe siècle av. J.-C. à la fin du Haut-Empire, 1 Les monuments publics, Picard, Paris. 503 p

Helly B (2013) Evolution topographique de Vienne du Ier au Ve siècles apr. J. C. In: Carte Archéologique de la Gaule, 38/3, Vienne. Académie des Inscriptions et Belles Lettres, Paris, pp 116–148

Helly B (2017) Vienne Antique. Sainte-Colombe, Saint-Romain-en-Gal, Vienne, Guides Archéologiques de la France, Éditions du Patrimoine, Paris, p 153

Helly B (2018) Sainte-Colombe (Rhône):11 rue du Docteur Trenel, Rapport de fouilled'archéologie préventive; Analyse d'un suburbium de Vienne antique; 2002–2018. Lyon, SRA Rhône-Alpes, p 46

Helly B, Rideaud A (2016) Ten years of archaeoseismological research in Armenia (2004–2013). Quat Int 395(2016):216–232

Lambert J (1985) Eléments de sismicité à Vienne au cinquième siècle (Isère), mise au point. Étude effectuée pour le commissariat atomique, BRGM. Note interne

Lucas G (2014) Sources textuelles gréco-latines sur Vienne, in Carte Archéologique de laGaule, 38/3, Vienne, Académie des Inscriptions et Belles Lettres, Paris

Reynaud J-F, Blain S, Delomier CH, Flammin A, Guibert P, Helly BE, Lanos PH (2019) La basilique de Saint-Ferréol et l'église paroissiale de Saint-Romain-en-Gal (Rhône). Revue archéologique de Narbonnaise 52:293–312

Rodríguez-Pascua MA, Pérez-López R, Giner-Robles JL, Silva PG, Garduño-Monroy VH, Reicherter K (2011) A comprehensive classification of Earthquake Archaeological Effects (EAE) in archaeoseismology: application to ancient remains of Roman and Mesoamerican cultures. Quat Int 242:20–30

Assessment and Reduction of Seismic Damage to Buildings of Historical Interest

11

The Reconstruction of Città di Castello—Umbria—after the Severe Earthquake of 1789

Francesco Giovanetti

Abstract

Today in Italy, interventions on listed buildings are regulated by Guidelines issued by the Presidency of the Council of Ministers on 9 February 2011 (Direttiva del Presidente del Consiglio dei Ministri 9 febbraio 2011. *Valutazione e riduzione del rischio sismico del patrimonio culturale con riferimento alle Norme tecniche per le costruzioni di cui al D.M. 14/01/2008* (pubblicata nella Gazzetta Ufficiale n. 47 del 26/02/2011—supplemento ordinario n. 54)), entitled "Assessment and Reduction of Seismic Damage to Buildings of Historical Interest". These Guidelines are the result of three decades worth of Italian research in the development of analytical methods to achieve an understanding of the seismic activity on historic buildings. For example, the huge innovation allowing verification from the local modes of damage is now introduced by current legislation.

Over the last three decades, a new interest in historical building techniques has spread widely, in recognition not only of their intrinsic value as a legacy of past material culture but also of their anti-seismic efficacy, which had not been taken seriously into consideration beforehand.

The pre-modern building, both ruins and living monuments, that houses a business requires a holistic analysis that combines the material substance of bricks and mortar, the architectural form, and its cultural significance.

Structural work with the task of reducing seismic vulnerability must be evaluated in the overall framework of the conservation of the building. The choice of strategies and intervention techniques, as well as the urgency to implement them, should be the consequence of an earlier evaluation phase. The main objective should always be the conservation, not only of the materials but also of the structural elements, in cases where no major deficiencies are present.

Keywords

Architectural heritage · Archival research · Historical earthquakes · Seismic prevention · Pre-modern building techniques · Repair of building, either destroyed, to be lowered or demolished · Calculate the cost of the demolition work, of the renovation of the building or necessary repairs · Theoretical modelling

F. Giovanetti (✉)
Dipartimento di Architettura e Restauro, Università di ROMA TRE, Rome, Italy

L. Pecchioli (ed.), *Archaeoseismology*, Natural Science in Archaeology,
https://doi.org/10.1007/978-3-031-28303-1_11

11.1 Chronology of the Seismic Events

A structural calculation, using analytically formulated theoretical models, represents a useful tool to evaluate building safety in the case of a historical building only on the condition that the mechanical model derives from a correct analysis of its mechanical behaviour, and that the designer is aware of the reliability limits of the results.

As a matter of fact, the values obtained from the structural modelling calculations are affected by intrinsic uncertainties both in the input data and in the model: the automatic calculation, however sophisticated it may be, does not ensure the reliability of the output. The choice of the designer should therefore give precedence to works that fit to integrate with the existing structure as well as avoiding radical transformations, that can change the original structural behaviour. As far as possible, the process should be reversible, non-invasive, and respectful of traditional techniques and materials.

Apart from the structural calculation according to modern methods, a qualitative analysis is decisive. The inner connection of masonry elements (Fig. 11.1), the geometry of the walls, the presence and quality of structural ties, and, last but not least, the possible thrusts coming from adjoining buildings all contribute to the structural characteristics of a historical building.

The study of earthquakes, that occurred before the seismic phenomena took on a scientific character, offers interesting insights for the conservation of the historical heritage built in masonry. A key example is a catastrophic tremor and tectonic earthquake that struck Città di Castello and its surroundings from the north-west in the afternoon of 30 September 1789, with an intensity estimated between 7–9 degrees on the Mercalli scale.

The Municipal Library and the Historical Archives of Città di Castello hold several important documents dating back to the months following the earthquake.

The paper presents the chronology of the seismic events and the assessment of the damage suffered by the most important buildings. It also considers the work carried out by the commission of architects sent to the site by Pope Pius VI for immediate works on the collapsing buildings and the technical prescriptions for reconstructing the city and the countryside.

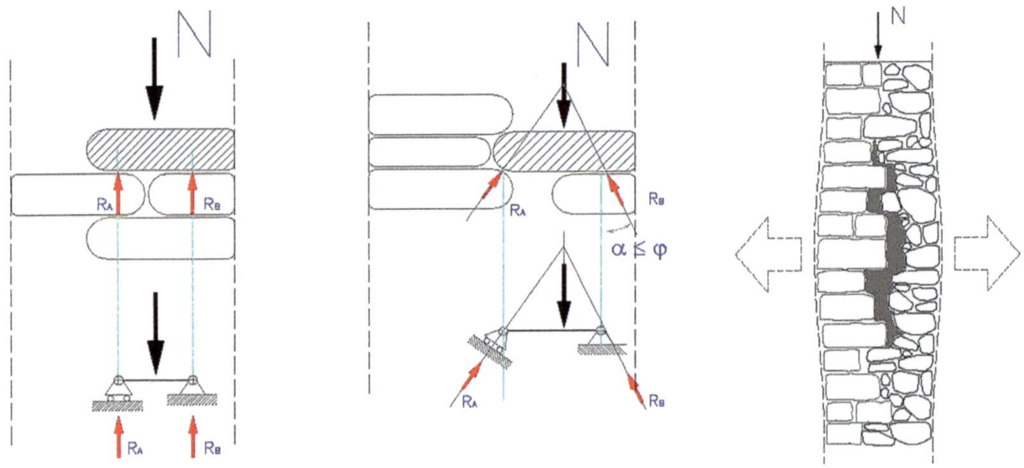

Equilibrium, due to the good meshing of the elements between the two faces of the wall.

Poorly meshed elements may generate lateral thrusts inside the wall.

Instability of the wall caused by internal lateral thrusts.

(courtesy Giovanni Cangi)

Fig. 11.1 Source: Cangi G. (2005) *Manuale del recupero strutturale e antisismico*, (Ed.) DEI, Roma

Two documents, although not dated, are to be placed in the first half of 1790 and represent a particular interest in our understanding of historical earthquakes.

The first is a text entitled, *A Faithful Account of the Terrible Earthquake in Città di Castello on the Morning of 30 September 1789.* The text consists of a detailed description of the event, offering an all-around picture of the effects of the earthquake on the buildings and the population (Fig. 11.2).

It provides important information on the major architectural and artistic buildings by describing the type and extent of the damage suffered by churches, palaces, convents, and institutional buildings. The damage to sacred images and other important artworks, frescoes and paintings is also listed. Furthermore, the effects on the population and the initiatives undertaken by authorities to tackle what we now call the *emergency* are described.

The printed sheets were published on the initiative of the local Jesuit College by the printing house of Fedele Toppi (active between 1777 and 1798). The author is not mentioned, but it is most likely Luigi Brami (1753–1818), a cleric and a scholar in the local *Accademia dei Liberi*.

The second is a group of manuscript papers entitled: *An Examination of the Damage Caused by the Earthquake of 30 September 1789 in Città di Castello*. Dated 29 April 1790, this document provides a detailed technical analysis of the damaged buildings in the city and in the countryside, including to ordinary houses.

Città di Castello in the province of Umbria, was one of the Papal States, so, shortly after the earthquake, the Apostolic Chamber sent the General Treasurer Monsignor Rufo from Rome to evaluate the damage and assess the cost of the repairs. He was aided by architects Andrea De Dominicis and Ferdinando Folcari from the Apostolic Chamber. On 13 October, he asked a third architect, Pietro Ferrari, the head of the Chamber of Commerce for the Province of Umbria and the Duchy of Spoleto, to attend.

Each of the three architects was assigned a sector of the city, which also comprised a corresponding sector outside the city wall. Each of them proceeded with inspections to assess which buildings should be demolished, in whole or part (most upper floors were beyond repair), and which ones could be preserved. Their task was also to immediately implement the provisional works necessary to prop up what could be repaired and, finally, to compile a detailed estimate of the costs.

The need to reduce the danger of collapsing buildings was urgent. Materials were stockpiled to be reused for future reconstruction and timber from collapsed buildings used for shoring was recovered. On 31 October, the General Treasurer ordered supplies of materials useful for securing works from the city of Ancona: "4 barrels of long nails; 6 barrels of large nails, to be used for fixings for wooden beams and other roof timbers; and 4,000 lbs of iron rods to connect floor and roof beams".

From this technical mission, only Pietro Ferrari's final report dated 29 April 1790 is known to have survived. It contains a description of all the damage to the houses in his sector as well as a calculation of the cost of repairs. A map showing the location of the buildings referred to in the text can no longer be found in the archives.

In submitting his repair estimate to the General Treasurer (over 1.3 million *scudi*—about € 40 billion today), Ferrari, although not requested, additionally proffered his technical advice on how to carry out repairs: "The circumstances of my observations have indicated the most appropriate means to achieve the firmest, the most stable, and the safest shelter to each building. Therefore, I also believe it is my office's duty to briefly mention the main actions to be taken for greater clarity, to aid those who will carry out the necessary repairs" (Figs. 11.3, 11.4, 11.5, and 11.6).

The importance of Ferrari's report lies in the keen technical directions he provides on the reconstruction works.

GENUINA, E DISTINTA

RELAZIONE

Dell' orribilissimo Terremoto

SCOPPIATO

IN CITTA' DI CASTELLO

La mattina dei 30. Settembre 1789.

DATA IN LUCE

PER PARTE DEL PUBBLICO DI DETTA CITTA';

A scanso dei molti equivoci, che si sono veduti in altre Relazioni uscite finora.

L' ultimo giorno dello scaduto Settembre 1789. ricorrendo la festiva memoria di *S. Girolamo* Dottor Massimo della Chiesa, comechè l'aria tranquilla, e il sereno Cielo non presagissero alcun fenomo luttuoso, questa Città poco dopo le ore 17. balzata venne per succussione, e ondulazione da un Terremoto così gagliardo, che nello spazio di circa venti minuti-secondi furono diroccati tutti gli edifizj, o guasti per gran maniera, e renduti inabitabili. L'antichissima Cattedral Chiesa, celebre fra le più belle d' Italia, che fin dall' anno 1503. fu dai fondamenti ridotta in miglior forma nello spazio di anni trenta coll' architettura di *Bramante Lazzari*, e coi disegni di *Raffael d'Urbino* di lui scolare, e che oltre i recenti ornati attualmente si riduceva all'ultima eleganza e perfezione, ha sofferto immensi danni per ogni parte. La gran cupola di detto Tempio arricchita di vaghe pitture a fresco del *Cavalier Mazzanti* rinomato, come ognun ben sa, per la vivezza del colorito, è giù piombata, ed ha sfondato il forte sotterraneo, ricoprendo colle macerie il venerato Sepolcro de' Santi Protettori *Florido*, ed *Amanzio*. Non ha però che lievemente patito la Tribuna fregiata dal pennello del Cavalier *Marco Benefialli*. E' caduta pur la cupola con laceramento di muri nella ben architettata Chiesa, e adorna di stucchi, Santuario frequentatissimo; della *Vergine di Belvedere*, un miglio e mezzo fuori della Città. Le altre Chiese, massimamente le Parrocchiali, sono sì malconce, che quasi tutte abbisognano di riedificazione. La Chiesa di *S. Sebastiano*, che riconosce per Protettore l' Emo *Cardinal Nipote*, e che di presente è stata dai Sigg. Canonici prescelta per la recita de' Divini Offizj, il nuovo magnifico Spedale-Unito degl' Infermi, Orfani, e Projetti, dove si lavorano i Calancà, Dobletti, ed altre tele di Cotone, il qual Luogo pio si sostiene

Fig. 11.2 The first page of the 1790 printed booklet describing the earthquake (Note 3)

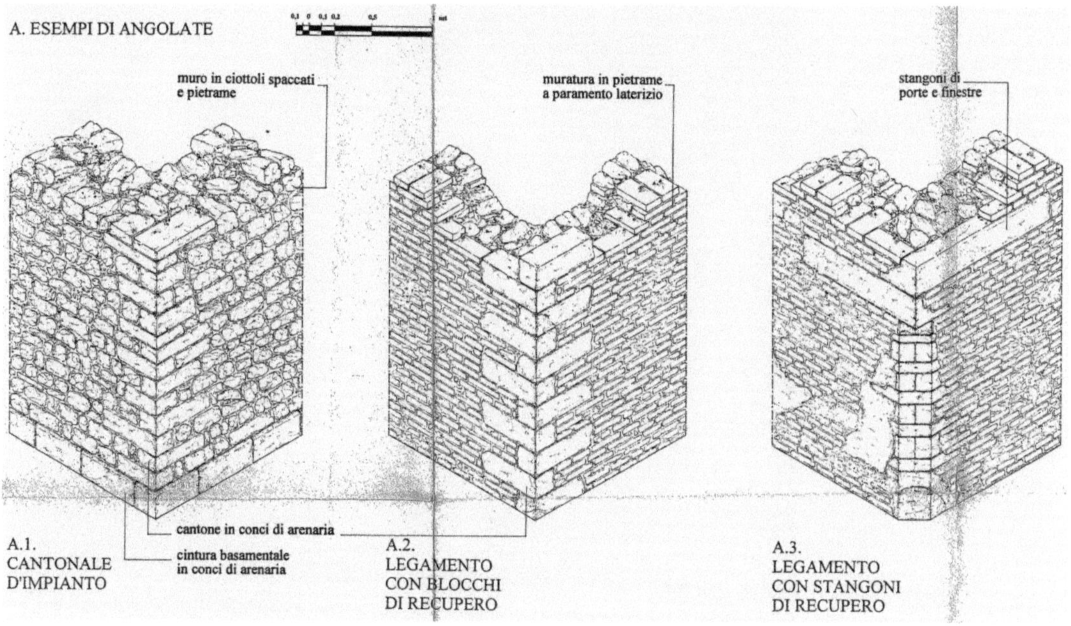

Fig. 11.3 After the earthquakes of 1353 and 1458 in Città di Castello, particular attention was given to the corner joints of the walls, often resorting to the reuse of stone elements taken from the ruins of the collapsed houses. A: *via del Borghesi*. The corner of the old church of S. Giovanni in Campo in sandstone blocks (twelfth century) is loosely joined to the masonry of stone and split pebbles. B: *via del Soccorso*. The corner (fifteenth century) is strengthened by a heavy joint of reused blocks deeply set in the masonry construction. C: *via delle Legne*. The bevelled corner (fourteenth–fifteenth centuries) consists of long masonry elements salvaged from the ruins of doors and window architraves

11.2 Città di Castello Recovery Handbook

The author consulted these papers in 1990 while preparing the second edition of the *Città di Castello Recovery Handbook*, published in 1992. A section of the Handbook was dedicated to traditional anti-seismic building techniques, in which he chose case studies that fitted with Pietro Ferrari's suggestions.

The fieldwork was supported by documentary research carried out in Città di Castello and the examination was extended to other seismic prevention techniques concerning the arches of the façades (Figs. 11.7, 11.8, and 11.9), the vaults inside the buildings, and the contrasting arches widely used to connect most of the city's buildings (Figs. 11.10, 11.11, 11.12, and 11.13).

Later, in 1993, the material collected, together with some additional drawings, was used for a study under the Direction of Paolo Marconi (1933–2013), then Professor of Architectural Restoration at the Faculty of Architecture, University of Rome. [1]

The documents consulted in 1992 appear to be greatly relevant today. From Ferrari's text, the direct, clear and intuitive criteria that guide his work emerge. His expertise lies in knowing how to "see" the construction, guided by knowledge informed by reason and intuition, acquired, as he writes, "from having put into practice the rules,

[1] On behalf of the Ministry for the Cultural and Environmental Heritage—Committee for Seismic Risk: Ministry for B.C.A. (Cultural Heritage)—Committee for the seismic risk. University of Rome "La Sapienza"—Chair of Architectural Restoration 1—Line 4.1.1—Historical seismic prevention techniques—Città di Castello. Direction: prof. P. Marconi: coordination by architect A. Pugliano; research: arch. F. Giovanetti. Arch. G. Boni, ing. G. Cangi, Rome 1993, (see Figs. 11.2, 11.4, 11.5, 11.6, 11.7, 11.9, 11.13, 11.14, and 11.15).

Fig. 11.4 Fourteenth-century bars reused as a strong connection of the corner (via delle Legne/via del Rialto). See Fig. 11.1c

the illumination, and all the knowledge, given to me by my profession". In other words, and today's language, we are subjecting the building to a qualitative analysis aided by experience.

* * *

Document no.1 [2]

A Faithful Account of the Terrible Earthquake in Città di Castello on the Morning of 30 September 1789. Drawn up to Illuminate the Public of that City and correct the many Misunderstandings of other Reports given Heretofore [3]

On the last day of the past September 1789, while celebrating the festive memory of St. Jerome [...] though tranquil air and the serene sky did not presage any mournful phenomenon, this city shortly after 5 pm, began jolting

[2] The square brackets indicate parts of the full text omitted for brevity.

[3] Città di Castello, *Municipal Library*, 4 printed sheets.

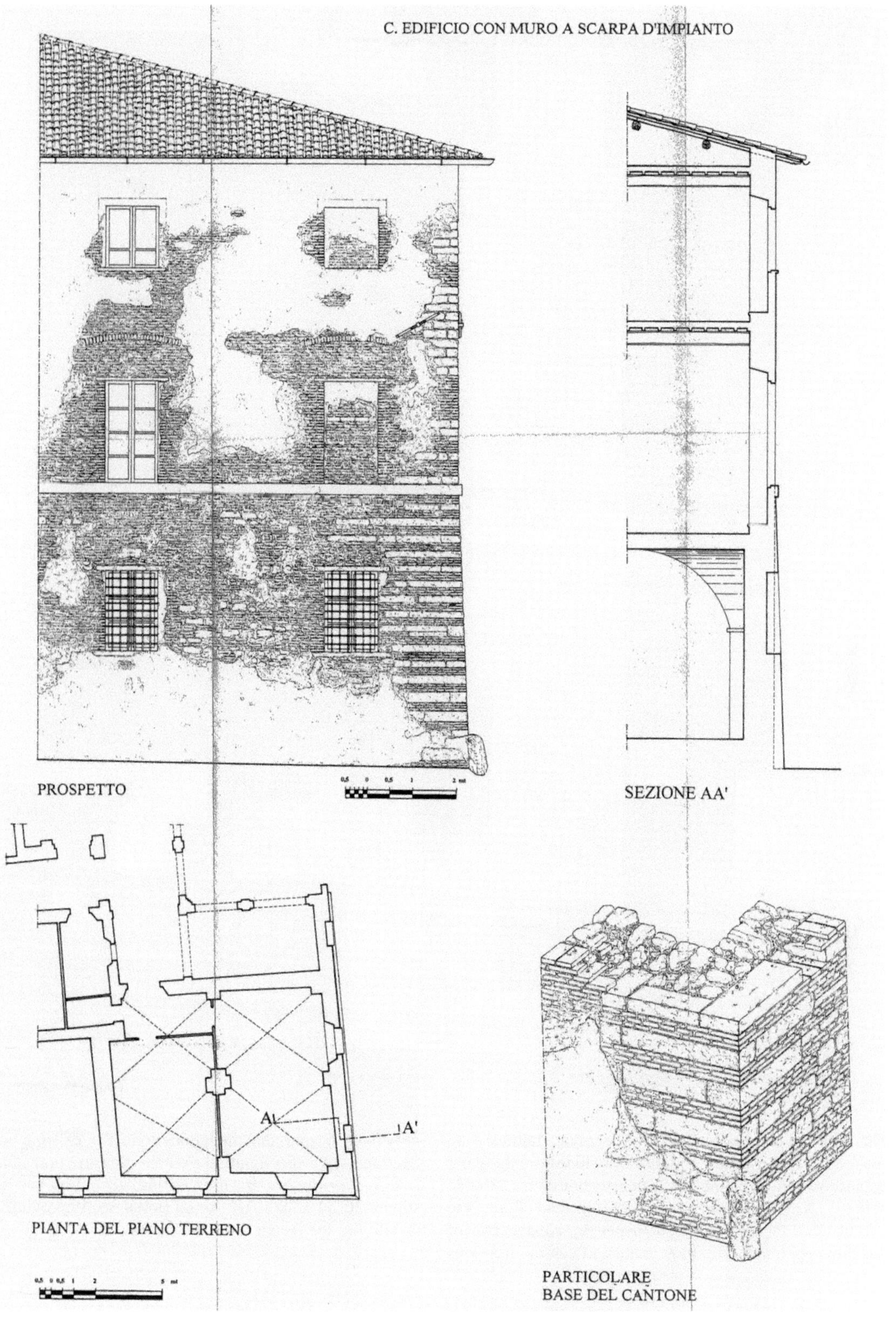

Fig. 11.5 The reinforcing of the older walls, as well as newly built buttresses was widely used after the earthquake of 1789, in accordance with the directions of the architect Ferrari. In *via dei Libelli*, the Vivarelli Colonna Palazzo refurbishment consisted of an inclined wall up to the first floor. Typical post-1789 masonry can be observed in the splayed corner: brick courses alternating with sandstone blocks salvaged from the ruins

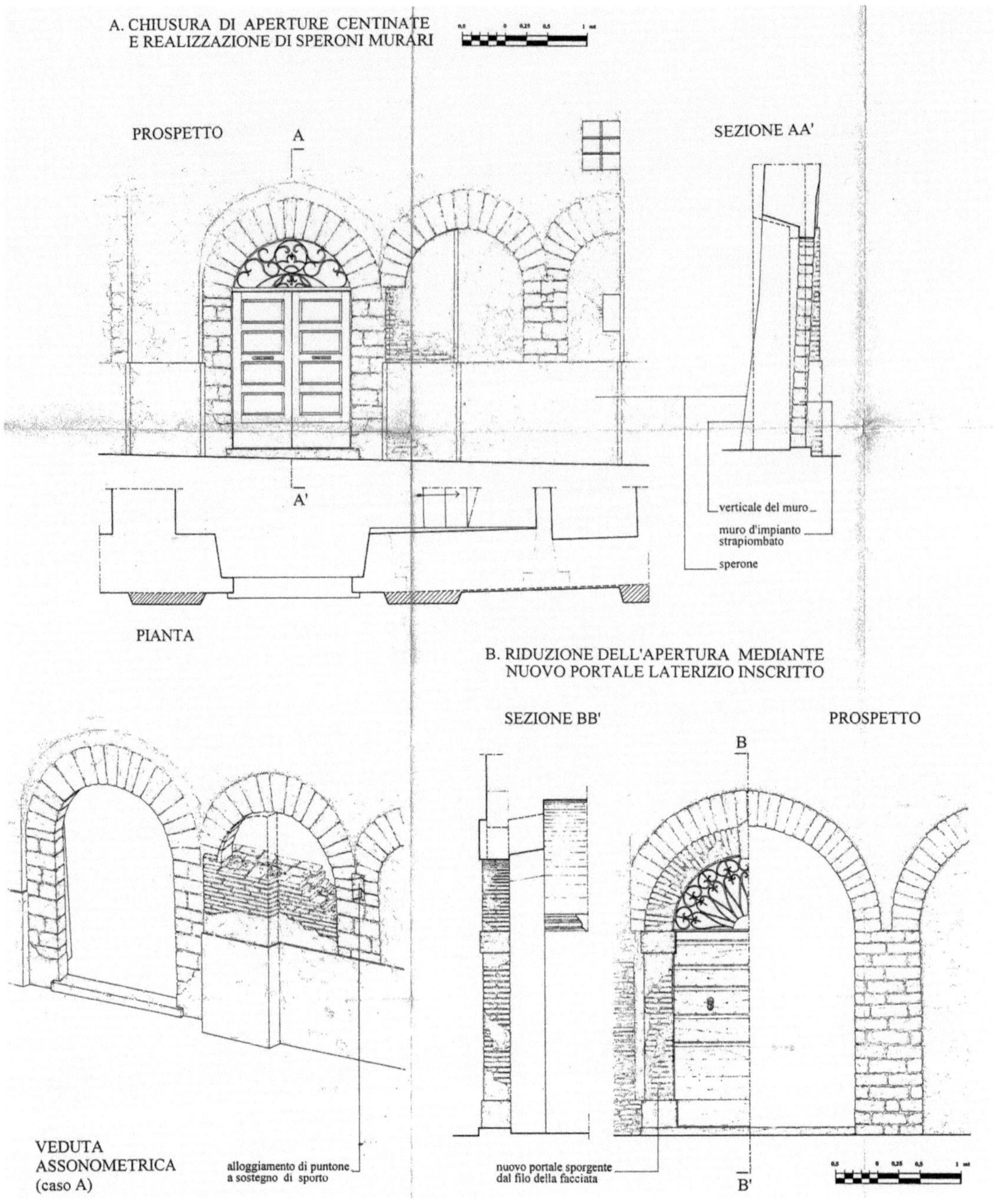

Fig. 11.6 In pre-modern buildings, applying buttresses to the façades was the most common way to restore the static balance of out-of-plumb walls due to transverse seismic actions. A.: in the loggia on the ground floor, the overhanging façade was strengthened by closing two of the three openings. The most damaged central arch was closed by a spur that integrates with the existing wall. Section A-A': the figure shows the construction phases *(via S. Florido 24)*. B.: the existing stone arch has been supported by a new 18th brick portal so, improving the stability of the façade *(via Plinio il Giovane 10)*

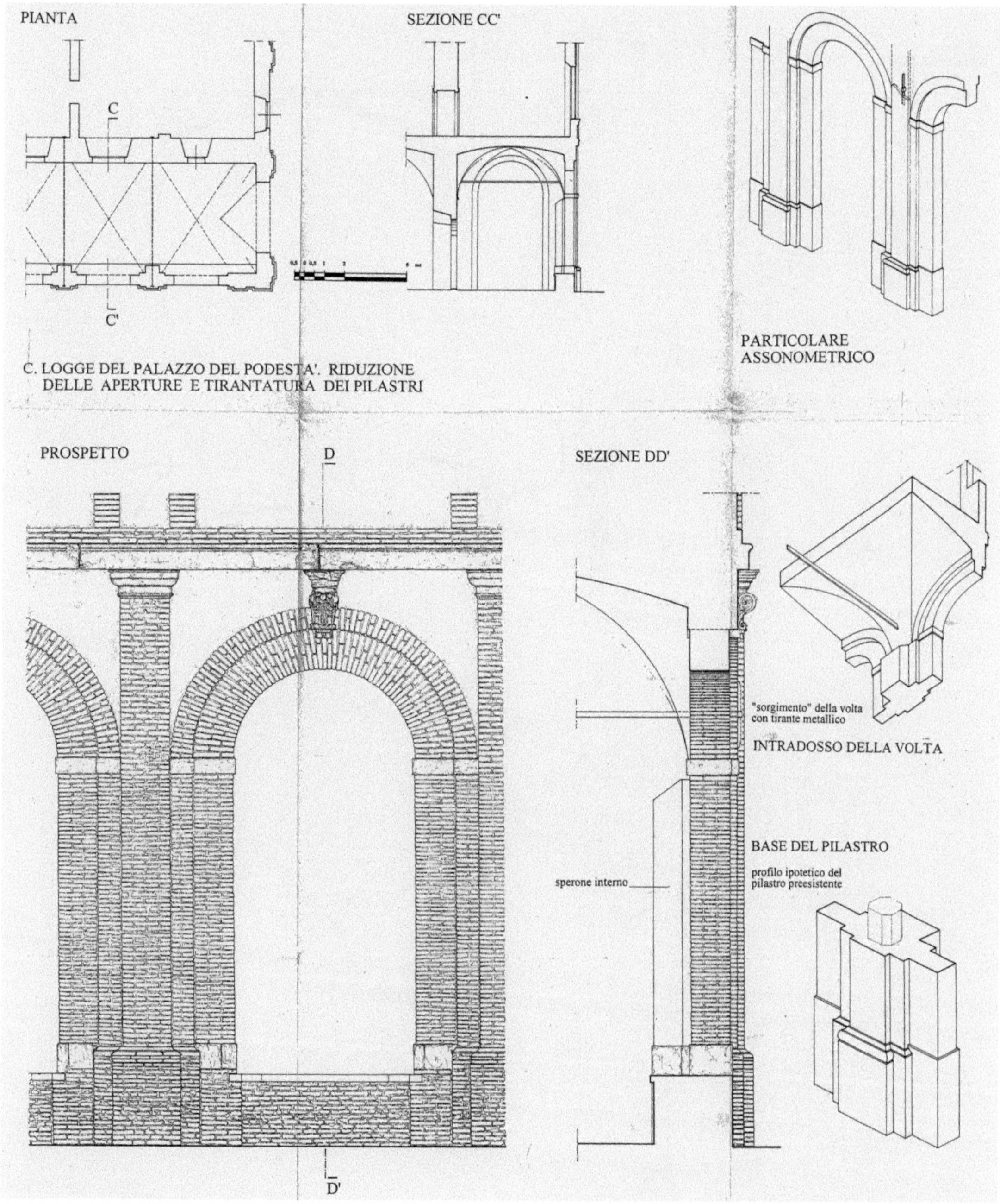

Fig. 11.7 The mediaeval portico of the Palazzo del Podestà *(piazza Fanti)* was extended and raised in the eighteenth century. The slender mediaeval pillars were incorporated into sturdy brick pillars secured to the rear main wall by tie rods. In contrast, new supporting arches were built on the intrados of the existing ones to improve longitudinal stability

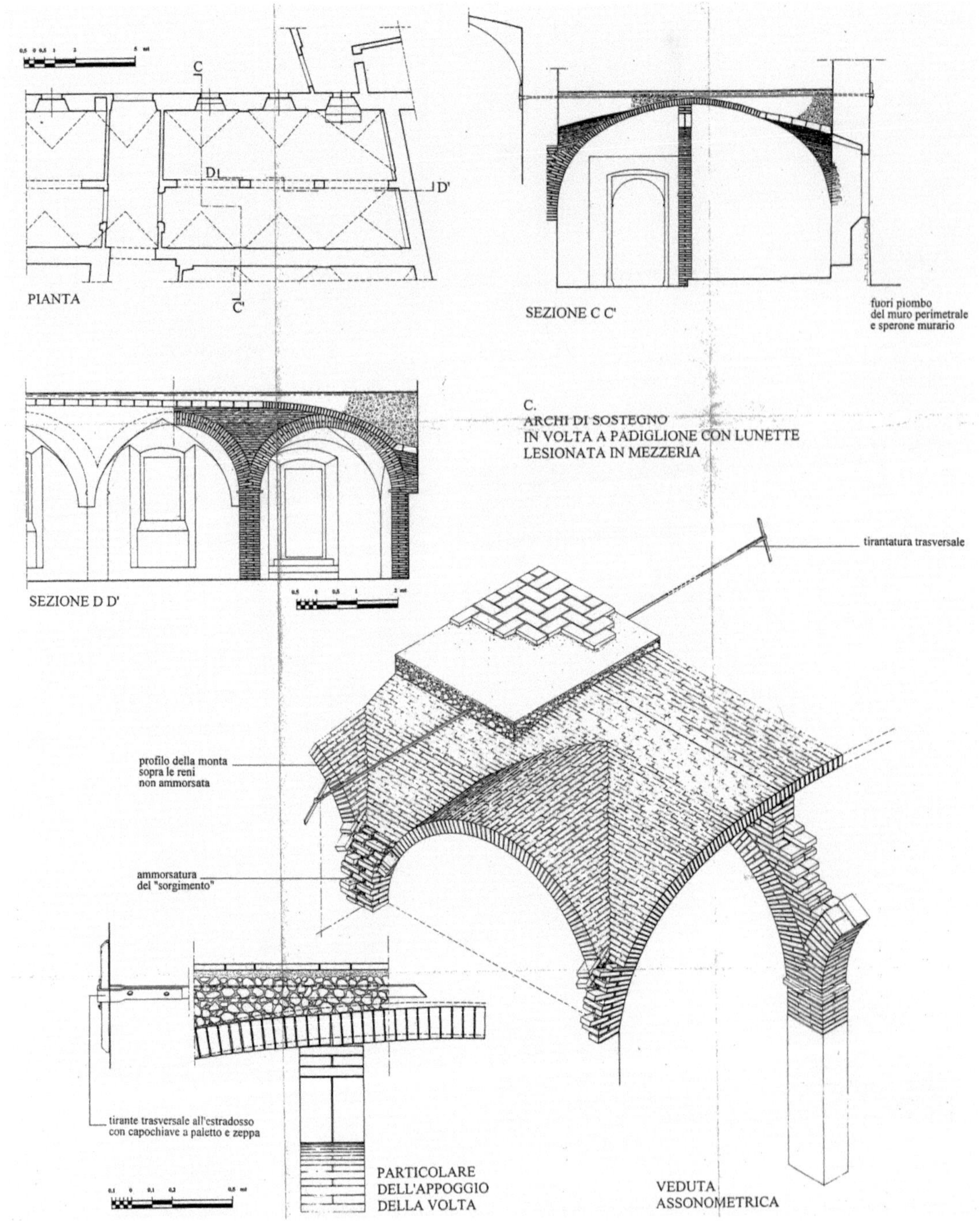

Fig. 11.8 In the Walled Clarisse nunnery *(via dei Lanari 2)*, the large pavilion vault with lunettes were damaged next to the centre line due to the mutual distancing of the supporting walls. A sequence of arches parallel to the longitudinal walls was placed to support the vault where the lesion occurred. The repair work is accompanied by transverse tie rods embedded in the extrados floor of the vault. Finally, the overhang of the external wall, pushed by the structural kinematics that triggered the lesions, was stabilized by a buttress

Fig. 11.9 The ribbed cross vault of a large mediaeval room in the ground floor of a Perugian merchant palace was divided by a wall, which later on also needed two robust arches to support the walls of the upper floors

and undulating with an earthquake so vigorous, that in the space of twenty seconds all the buildings were destroyed, or utterly ruined, and rendered uninhabitable. The most ancient Cathedral Church, celebrated amongst the most beautiful in Italy, which from the year 1503 was built up from its foundations by Bramante Lazzari into its better form over thirty years, and with the decoration of Rafael of Urbino his pupil, [...] has suffered immense damage in every part: The great dome of the said Temple [...] has plummeted asunder breaking through the robust crypt, and covering the venerated tomb of the Patron Saints Florido and Amanzio with debris. The Tribune decorated by the brush of Cavalier Marco Benefialli has not suffered so much. A mile and a half outside the City, in the beautifully designed and adorned Church of the Virgin of Belvedere, a much frequented sanctuary, the dome has also collapsed tearing the walls down with it.

The other churches, especially the Parish churches, are indeed in a sorry state, and almost all require rebuilding. The Church of San Sebastiano, [...], the new magnificent Unified Hospital of the Sick, Orphans and Abandoned children, where the fine handkerchiefs, French tissues, and other cotton fabrics are worked [...] were found the least affected by the universal punishment. Only minor detriment was cast on the new Theater, erected with ingenious and elegant design, [4] *where many families without shelter have taken refuge.*

[4] Built by the roman architect Girolamo Masi, who published a treaty in the previous year: *Teoria e pratica di architettura civile per istruzione della gioventù specialmente romana*, (Fulgoni, Roma 1788).

Fig. 11.10 Città di Castello *(dead end from via Cavour)*: pushing arches called "archi di sbatacchio" intended to connect the decks of opposite buildings overlooking the city streets, with the task of synchronizing seismic oscillations, thus avoiding the sliding of the beams and disruption of the vaults

The Vitelli palace, once decorated by Niccolò delle Pomarance, Prospero Fontana, Orazio Sammachini, and by other talented artists, once a wonder of this city and the erudite entertainment of curious strangers, is destroyed internally by the collapse of its heavy vaults, leaving the external walls standing but broken and crooked. The water which was carried to the garden in abundance by underground channels of the said palace, and springing into the air via the many magnificent fountains, lakes, and grotesques, giving a beautiful ornament to the garden, at present, it is totally diverted such that the said fountains have remained completely dry, the water no longer leading nowhere.

Of the Town Hall and the palaces, namely, the Apostolic, the Episcopal, Bufalini, Ubaldini, and

Fig. 11.11 Venice, thrusting arches: to improve the connection between two floors placed at different heights and the distance between the buildings an inverted arch had been built for better resistance to seismic oscillations

especially Berioli, among others, we see nothing but ruin, or threat of near collapse.

Without going into macabre detail, I would report the miserable condition of the other buildings of this city are in, which, although small, was one of the most magnificent and beautiful cities of the Papal States for a host of reasons: for the splendour of the temples, for the grandeur of the buildings, for the convenience of the streets, and the regularity of the squares.

The great shock of the earthquake began from the side where the Tramontana wind blows [north] towards S. Sepolcro. At the impetuous blast, the land cracked in large openings from place to place and towards the Villa di Selci four miles distant from us. Pools of water gushed out, which shortly after dried up, depositing much deep green sand, mixed with some particles of iron, nitro, and sulphur. Along that line the best homes of the countryside were razed to the ground. In the Villa del Bagno about a mile from the city, once a source of sulphurous water, which for some time had almost dried up, has now reemerged in such a flood that it would make a grindstone turn.

In the aforementioned Villa of Selci, forty people died, in the villas of Grumale and Cerbara

Fig. 11.12 Città di Castello, via Cacciatori del Tevere: an earthquake-damaged attic has been supplemented by masonry arches

hardly a stone is standing, likewise where the palace of the Counts Berioli once stood.

In the populated village of Villa of S. Giustino, beyond the other destroyed houses, the ancient military garrison with its devastated interior can be seen, along with the Summer Palace of the noble Bufalini Family, which in the autumn months was a place dedicated to gracious hospitality.

The pleasant country houses of the Marquis Vitelli, Count Ubaldini, the Graziani family, and other noble families have met the same disastrous fate. Many unfortunate folks were forced to stay out in the open for several days, suffering the season's nocturnal cold and long intolerable downpours. The huts and shacks, even the stables serve as shelter for those who were accustomed to live in the most affluent well-appointed lodgings.

And yet they count themselves lucky, if they have lost all else, in having saved their lives.

Ten people died in the City, and less than a hundred died in the Countryside; and twixt the City Country, there are approximately one hundred wounded. It is a miracle that so many survived the terrible scourge.

[. . .]

The Cardinal Braschi-Onesti [on behalf of the Pope] sent as Emissary one of this wisest Ministers, His Most Reverend Excellency Monsignor Ruffo Treasurer General, so that the immense damage would be examined, in the company of the most expert Mr. Andrea De Dominicis, Mr. Ferdinando Folcari, and Mr. Pietro Ferrari Architect of the Most Reverend Chamber, and would disburse a considerable sum of money, and much more would promise, for the betterment of the miserable and desolate populace.

[Furthermore, he. . .] visited the whole of the remaining City over and over again, [. . .] exposing himself to imminent danger without fear. To remove the risk of theft, and to maintain good order, he approved the service of the urban Militia, and commanded that, in addition to the Corsican Guards, the Finance Guard should also be armed. [. . .] he gave everyone consolation. All were fully satisfied, and astonished by his charitable, indeed caring, and prudent Character.

[. . .]

A large number of builders have already been put to work and others are called from every part. Walls in danger of collapse have so far been safely lowered or propped. Roofs have been covered with the greatest solicitude. Afterwards, the reconstruction, the almost total renovation of Città di Castello will begin.

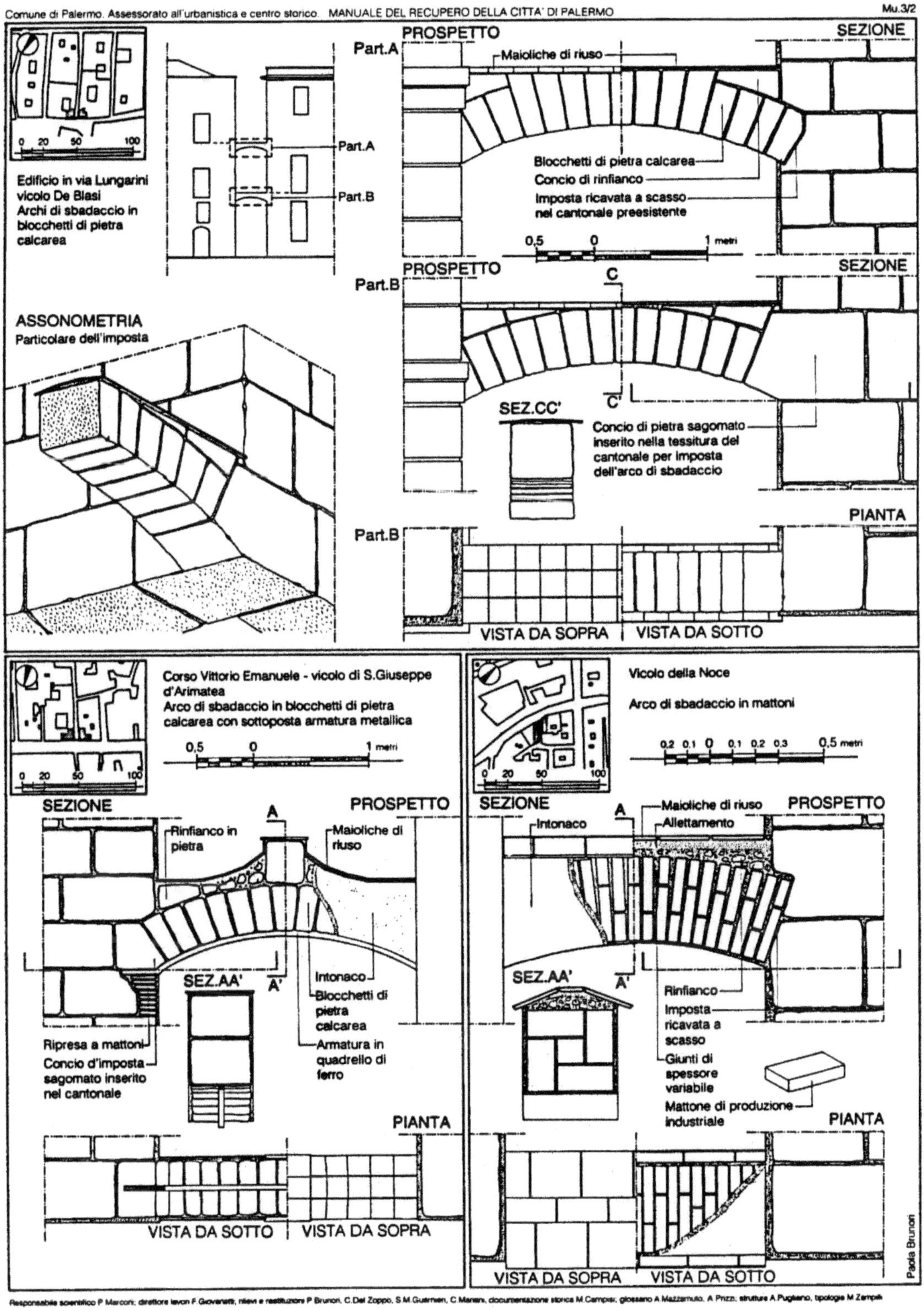

Fig. 11.13 Palermo: different types of anti-seismic arches built mainly of sandstone blocks. F. Giovanetti (ed.): *Manuale del recupero della città di Palermo*, cit.Tav. Mu 3/2

In the tifernate [of Città di Castello] seminar and college
care of Fedele Toppi
with permission

* * *

Document no.2 [5]

Committee for the allotment of subsidies for the damage caused by the earthquake of 30 September 1789. Città di Castello. Handwritten report by Pietro Ferrari, in the book concerning the examination of the damage caused by the earthquake of September 30, 1789, in Città di Castello, Rome, April 29, 1790. [6]

The most Munificent Prince his Holiness Our Lord Pope Pius VI [. . .] when the terrible news of the immense damage caused by the earthquake of 30 September last year 1789 [. . .], He deigned to send one of His wisest and vigilant Ministers, His Excellency Most Reverend Monsignor General Treasurer, so that he provides copious grants, to repair the damage and provide for the most urgent needs of those poor subjects. As soon as he arrived, he thought that a third architect was necessary in addition to the two already sent from Rome, to give prompt remedies to the buildings of the city and the countryside, which threatened impending collapse, and furthermore to examine the damage with the greatest solicitude caused by the earthquake in each House, designed [. . .] to [. . .] order me on my immediate arrival that I should visit the unsafe dwellings of a part of the city, and a portion of countryside to the south, to see to temporary shelter provisions for the many unfortunate families during the winter season, with shoring and demolishing works in those cases in which a major collapse was feared, as well as to assess the damage the buildings have suffered, and the costs necessary for the restoration, or the almost complete rebuilding necessary of the many houses to be re-inhabited.

In execution of these orders, I immediately got to work with great zeal, spurred by the urgency of the situation; at the same time, the other two architects were employed in other parts, in addition to the provisions taken, the orders given, and the assistance given to the propping, and provisional repairs of many buildings, I visited the houses described below one by one, I assessed the damages caused by the earthquake, and I calculated the cost of the repair.

[. . .]

Having therefore conducted my work based on the received instructions to examine the amount of damage to be repaired in both town and country buildings, I first had in mind the state in which the houses were before the earthquake to account only for the deterioration due to this cause.

Thus, following this rule, in those Houses in which the top floor was all or partly destroyed, or had to be demolished for its safety, I only calculated the cost of the demolition work, the renovation of the roof and necessary repairs, both on the ground floor and in the upper floor. In those, which had to be rebuilt from the ground, or from the ground floor upwards, I only calculated the amount for the reconstruction of the ground floor, and first upper floor. For houses with only a single room per floor belonging to poor people, I calculated a complete reconstruction with a first floor as well, necessary for a family dwelling, for which a single room that also serves as a kitchen can never be sufficient. As for the ruined churches, I calculated the amount of rebuilding according to their previous size. However, I only allowed for a simple but decent roof. Finally, in those buildings, which can be restored without losing a floor, I evaluated all the repairs that are needed on each floor. I must also advise, that in fixing the amount for the repair of each building, either totally destroyed, or to be lowered by a storey, or from demolition I considered the material that could be re-utilized. As for new materials, especially for the countryside buildings, I was mindful of the distance and therefore the costs of transport.

These are the precepts I followed in executing the task given to me. The circumstances of my observations have indicated to me the most

[5] The square brackets indicate parts of the full text eliminated for brevity.

[6] *Historical Archive of Città di Castello*, file to be numbered, temporary placement no. 3.234.

appropriate means to achieve the firmest, the most stable, and the safest shelter for each building. I, therefore, also believe it is beholden to my office to briefly mention the main actions to be taken for greater clarity, to aid those who will carry out the necessary repairs.

In buildings, therefore, in which the shock of the earthquake has not caused much damage, but only some minor loosening, and fissures in the walls, it will be useful to use iron chains to fix the weakest parts, and iron brackets to fix the head of the floor beams lodged in the outer walls; and where these beams meet, the use of brackets will always be better than that of iron chains, which, while standing in their stead and performing the same function, are stronger and prevent the beams hitting the walls in an earthquake. Experience tells us that iron chains tend to break in a violent shock (Figs. 11.14, 11.15, and 11.16).

In the more damaged houses, where the greater destruction is usually in the roof, the repair below can occur without removing the roof. It can be supported with props for the time necessary to repair the damaged part below to save the roofs: by far the most expensive part of the building and always the cause of great failures. The building that has suffered damage to the walls that are now out of plumb, the best, and the most effective means of restoring their previous stability and preserving them in time, the best way is that of a buttress, which having little space to extend into the narrower streets, must be constructed absorbing the old wall of the house, thus performing the same structural function and at the same time almost completely renovating the outer wall of the building without any deformity. Experience has shown me in many cases that buildings built with the walls sloping towards the street, are the least damaged by earthquakes, almost everywhere. It cannot be more highly commended that the external walls of the buildings be constructed from the foundations and raised with a steep inclination up to the first floor. In this way, they are stronger and more capable of withstanding both ordinary and accidental impacts over time.

After the external walls have been secured, either through new buttresses, or with the restoration of existing ones, the cracks of the internal walls must now be thoroughly filled, opening them up in sections two or three palms wide [roman palm: 22,34 cm] using good materials perfectly adapted to the wall, so as not to reappear at the slightest shock; since the stability of the renovation depends largely on the soundness of these bindings.

This is the result of the considerations, according to the aforementioned instructions, I made in assessing the damage and proposing the repairs of the houses, which have been visited within all the blocks. They were divided into four parts marked with numbers of different colours to avoid any confusion and facilitate recognition. In the description of these Houses I have indicated a lump sum amount for each restoration resulting from both the quantities and from my personal evaluation in exact accordance with the venerated orders of His Most Reverend Excellency.

[...] To achieve this end, I have put into practice the rules, the illumination, and all the knowledge, given to me by my profession. I have spared no diligence, care, or effort, as required in the service of the Prince, the debt of my office, and the importance of the task to be assigned.

* * *

The author, Architect, former director of the *Monumenti di Roma. Scavi, restauri e siti UNESCO*, within the Sovrintendenza ai beni culturali of the Rome City Council; former chair of *ARCo_Associazione per il recupero del costruito storico*; is a Professorial Teaching Fellow at the University of Roma Tre, Department of Architecture within the "International Master's level II program in Restauro architettonico e cultura del patrimonio".

This paper is a reworking of material which initially appeared in: F. Giovanetti (ed.), *Manuale del recupero del comune di Città di Castello*, Roma (DEI) 1992, pp. 20, 47 (ISBN 88.7722.577.7); then in: F. Giovanetti, *Un terremoto del 1789*, in "Ricerche di storia dell'arte", vol. 2009, pp. 54–63 (ISSN 03392-7202).

Rome,
January 18, 2021.

Fig. 11.14 E. Edge beam and rafters integrated with tie rods. F.1, F.2. Types of pre-modern metal tie rods. F.3, F.4. Diverse types of twentieth century metal tie rods

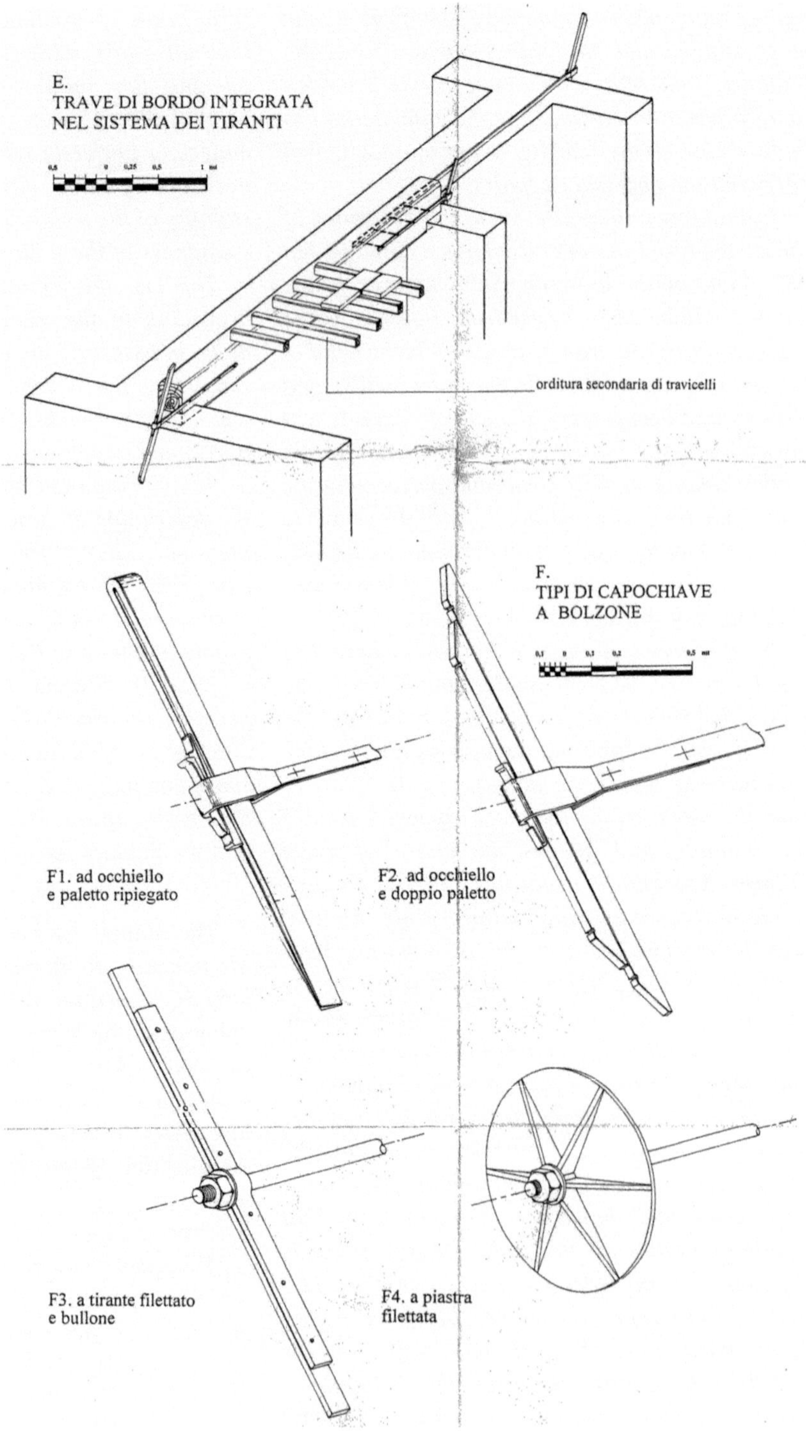

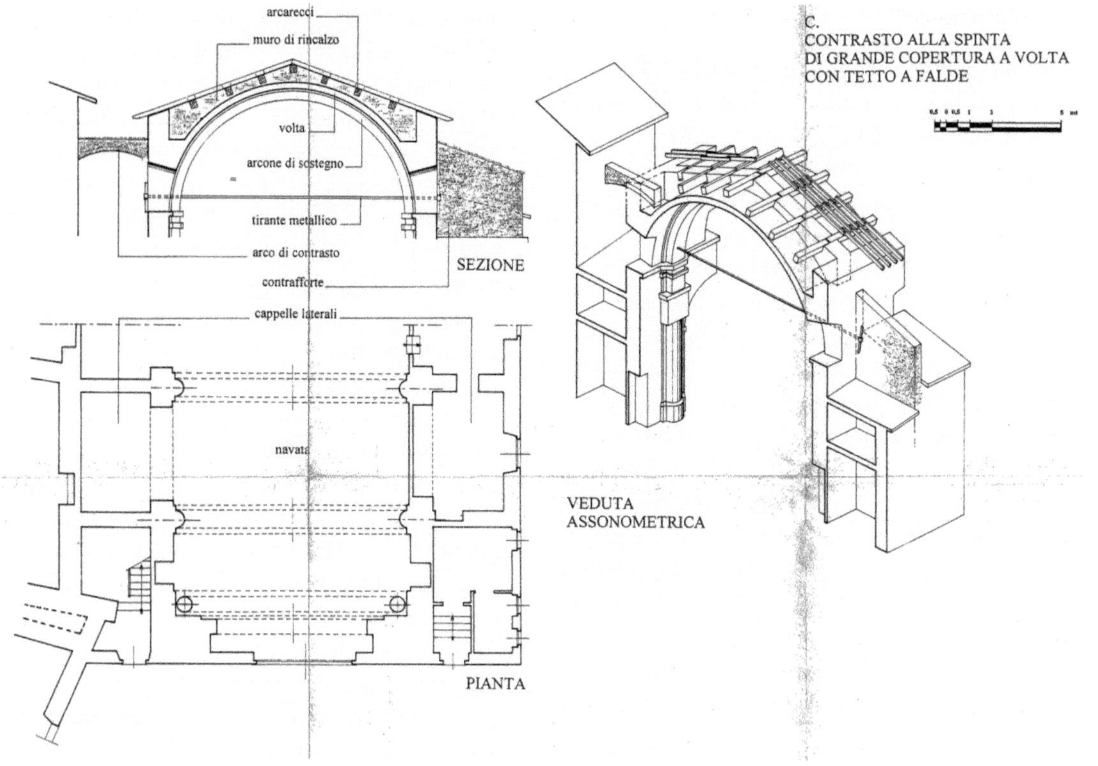

Fig. 11.15 The strong thrust of the Philippine Convent vault was reinforced by a thrust arch leaning against the left building, while internally a counter vault was built. The original body of the building is crossed by strong metal tie rods that connect to a thrust masonry

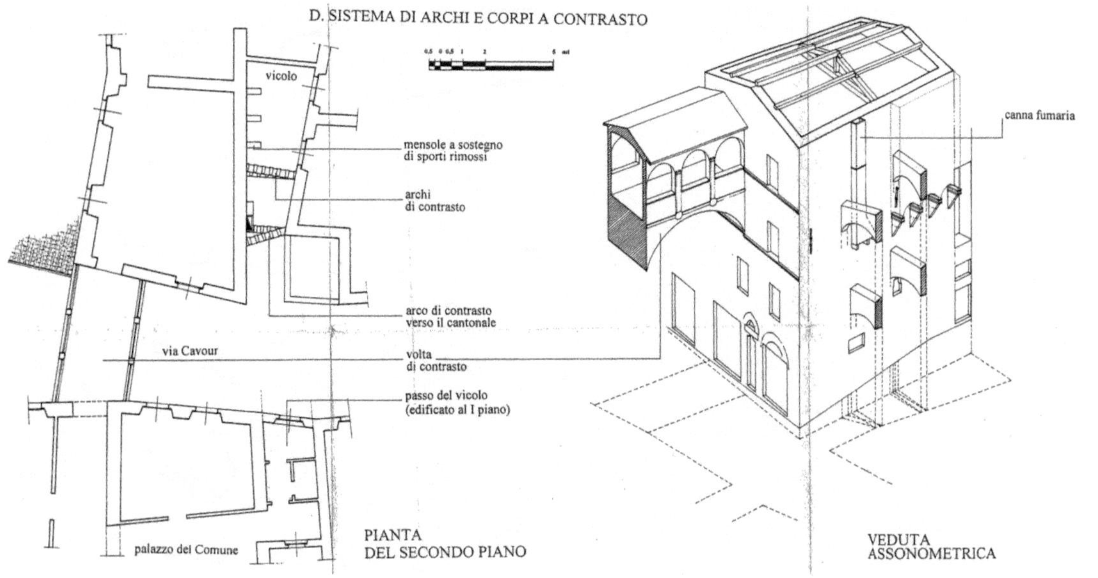

Fig. 11.16 Above Cavour Street, a contrasting vault is made between the City Hall building and that on the upper left. On the right, the axonometry highlights the new arches that replace the decommissioned mediaeval corbels

Detailed Notes

1. Municipality of Città di Castello, Perugia, Italy. Library Archives Service, via Largo Bartolini: zip 06012
   Phone: 0758529249; 3351203409
   E-mail: biblioteca@cittadicastello.gov.it
   Typology: territorial public body
   The archives held by the Municipality of Città di Castello are kept in various locations.
2. Pius VI, namely, Angelo Onofrio Melchiorre Natale Giovanni Antonio or Giannangelo Braschi; Born in Cesena, Dec. 25, 1717, 250th bishop of Rome, Pope of the Catholic Church from Feb 15, 1775. Died in Valence-sur-Rhône, Aug. 29, 1799.
3. The paper was published on the initiative of the local Jesuit College by the printing house of Fedele Toppi (active between 1777 and 1798). The author is not mentioned, but it is most likely Luigi Brami (1753–1818), a cleric and a scholar in the local *Accademia dei Liberi*.
4. The papers were first consulted in 1990 while preparing the second edition of the *Città di Castello Recovery Handbook*, published in 1992, with the advice of Antonino Giuffré (1933–1997), then a Professor of Architectural Restoration at the Faculty of Architecture, University of Rome "La Sapienza".
5. Later, in 1993, the material collected then, together with some additional drawings, was resumed for a study under the Direction of Paolo Marconi (1933–2013), then Full professor of Architectural Restoration at the Faculty of Architecture, University of Rome "La Sapienza", on behalf of the Ministry for the Cultural and Environmental Heritage—Committee for Seismic Risk.
6. Mention should be made of the new Theater *degli Illuminati* di Città di Castello, built by the Roman architect Girolamo Masi who had published—the very previous year—the important architectural treatise: *Theory and Practice of Civil Architecture for the Instruction of the Youth Especially of Rome (Fulgoni, Rome 1788)*. The building, still in use today, suffered little from the devastation that ravaged the city.
7. Soon after the earthquake, the Apostolic Chamber sent the General Treasurer Monsignor Ruffo from Rome to evaluate the damages and assess the cost of the repairs.
8. Fabrizio Ruffo (16.09.1744–13.12.1827). Enlightened papal dignitary and successful European politician, appointed in 1785 by Pius VI to the position of treasurer general. Due to the political controversies that beset him, he was dismissed by the Pope in 1794 and compensated by appointment as Cardinal.

Additional References

Ambraseys N (2005a) Archaeoseismology and neocatastrophism. Seismol Res Lett 76(5):560–564. https://doi.org/10.1785/gssrl.76.5.560

Ambraseys N (2005b) Historical earthquakes in Jerusalem – a methodological discussion. J Seismol 9(3):329–340

Ambraseys NN (2006) Earthquakes and archaeology. J Archaeol Sci 33(7):1008–1016

Arrighetti A (2022) Analisi e documentazione archeologica dell'edilizia storica in aree a rischio sismico: dall'atlante dei tipi costruttivi murari all'impiego delle moderne tecnologie di rilievo. Archeologia e Calcolatori 33.1., Firenze:201–218. https://doi.org/10.19282/ac.33.1.2022.11

Arrighetti A, Razzante V, Dessales H (2022) Archaeology and earthquakes in Siena (Italy). Preliminary data from the survey of the historical buildings of the Terzo di Città. Restauro Archeologico 1(2022):14–31. https://doi.org/10.36253/rar-13099

Avni R, Bowman D, Shapira A, Nur A (2002) Erroneous interpretation of historical documents related to the epicenter of the 1927 Jericho earthquake in the Holy Land. J Seismol 6(6):469–476

Bilham R (2009) The seismic future of cities. Bull Earthq Eng 7(4):839–887. https://doi.org/10.1007/s10518-009-9147-0

Bonilla MG (1992) Geologic and historical factors affecting earthquake damage. In: O'Rourke TD (ed) The Loma Prieta, California, Earthquake of October 17, 1989—Marina District. United States Government Printing Office, Washington, DC. F7–F34. U. S. Geological Survey Professional Paper 1551-F, Available online: http://pubs.usgs.gov/pp/pp1551/pp1551f/

Bottari C (2015) Archaeoseismology in Sicily: past earthquakes and effects on ancient society. In: D'Amico S (ed) Earthquakes and their Impact on Society. Springer, Cham, pp 491–504. https://doi.org/10.1007/978-3-319-21753-6_20

Caputo R, Helly B (2008) The use of distinct disciplines to investigate past earthquakes. Tectonophysics 453(1–4): 7–19. https://doi.org/10.1016/j.tecto.2007.05.007

Caputo R, Helly B, Pavlides S, Papadopoulos G (2006) Archaeo- and palaeoseismological investigations in northern Thessaly (Greece): insights for the seismic potential of the region. Nat Hazards 39(2):195–212. https://doi.org/10.1007/S11069-006-0023-9

Di Buccio F, Comina C, Fontana D, Minarelli L, Vagnon F, Amoroso S (2023) Fines content determination through geotechnical and geophysical tests for liquefaction assessment in the Emilia alluvial plain (Ferrara, Italy). Soil Dyn Earthq Eng 173:108057., ISSN 0267-7261. https://doi.org/10.1016/j.soildyn.2023.108057

Evans A (2013) The Palace of Minos: a comparative account of the successive stages of the early Cretan civilization as illustrated by the discoveries at Knossos (Vol. 4). Cambridge University Press (Original version: Evans, A.J. 1928. The Palace of Minos, vol. 2, London). https://doi.org/10.1017/CBO9781139795142

Fäh D, Steimen S, Oprsal I et al (2006) The earthquake of 250 A.D. in Augusta Raurica, A real event with a 3D site effect? J Seismol 10:459–477. https://doi.org/10.1007/s10950-006-9031-1

Fäh D, Gisler M, Jaggi B, Kästli P, Lutz T, Masciadri V, Matt C, Mayer-Rosa D, Rippmann D, Schwarz-Zanetti G, Tauber J, Wenk T (2009) The 1356 Basel earthquake: an interdisciplinary revision. Geophys J Int 178:351–374. https://doi.org/10.1111/j.1365-246X.2009.04130.x

Forlin P, Gerrard CM (2017) The archaeology of earthquakes: the application of adaptive cycles to seismically-affected communities in late medieval Europe. Quat Int 446:95. https://doi.org/10.1016/j.quaint.2017.06.030

Galadini G (2021) Le tre velocità. Percorsi della cultura del terremoto in Abruzzo tra scienza, normativa e proposta politica. Aleph Editrice, Luco dei Marsi, Italy, 198

Galadini F, Hinzen KG, Stiros S (2006) Archaeoseismology: methodological issues and procedure. J Seismol 10(4):395–414. https://doi.org/10.1007/s10950-006-9027-x

Galadini F, Ricci G, Falcucci E et al (2018) Archaeoseismological evidence of past earthquakes in Rome (fifth to ninth century A.D.) used to quantify dating uncertainties and coseismic damage. Nat

L. Pecchioli (ed.), *Archaeoseismology*, Natural Science in Archaeology,
https://doi.org/10.1007/978-3-031-28303-1

Hazards 94:319–348. https://doi.org/10.1007/s11069-018-3390-0

Galadini F, Ceccaroni E, Dixit Dominus G et al (2022) Combining earth sciences with archaeology to investigate natural risks related to the cultural heritage of the Marsica region (central Apennines, Italy). Mediterr Geosci Rev 4:287

Galli PAC, Molin D (2014) Beyond the damage threshold: the historic earthquakes of Rome. Bull Earthq Eng 12:1277–1306. https://doi.org/10.1007/s10518-012-9409-0

Godschalk DR (2003) Urban hazard mitigation: creating resilient cities. Nat Hazards Rev 4(3):136–143. https://doi.org/10.1061/(ASCE)1527-6988(2003)4:3(136)

Guidoboni E (a cura di) (1989) I terremoti prima del Mille in Italia e nell'area mediterranea. Storia Archeologia Sismologia, ING-SGA, Bologna, 766

Guidoboni E (1996) Archaeology and historical seismology: the need for collaboration in the Mediterranean area. In: Stiros S, Jones RE (eds) Archaeoseismology. The British School at Athens, Athens, pp 7–13

Guidoboni E (1997) An early project for an antiseismic house in Italy: Pirro Ligorio's manuscript treatise of 1570-74. Eur Earthq Eng 4:1–18

Guidoboni E (2000) Archeosismologia, voce in "Dizionario di Archeologia", a cura di R. Francovich e D. Manacorda, Laterza, Roma-Bari, pp 29–34

Guidoboni E (2016) Erdbeben und Seebeben im antiken Mittelmeerraum. Gründe für einen Dialog zwischen Seismologie und Geschichte, in "Erdbeben in der Antike", Deutungen – Folgen – Repräsentationen, herausgegeben von J. Borsch und L. Carrara, Mohr Siebec, Tubingen Univ., pp 15–26

Guidoboni E (2021) Sull'iconografia dei terremoti dal medioevo al Settecento. Memorie Descr. Carta Geol. d'Italia. vol. 108, a cura di ISPRA, 241–252

Guidoboni E, Comastri A (2015) The earthquakes of Aleppo and the region of "dead cities" (Syria) 7th–15th century. In: Riis T (ed) The dead cities of northern Syria and their demise, vol X. Publications of the Danish Institute of Damascus, Kiel, pp 61–110

Guidoboni E, Santoro Bianchi S (1995) Collapses and seismic collapses in archaeology: proposal for a thematic atlas. Ann Geofis 38(5–6):1013–1017

Guidoboni E, Valensise G (2015) On the complexity of earthquake sequences: a historical seismology perspective based on the L'Aquila seismicity (Abruzzo, Central Italy), 1315-1915. Earthq Struct 8(1):153–184. https://doi.org/10.12989/eas.2015.8.1.153

Guidoboni E, Muggia A, Marconi C, Boschi E (2002) A case study in archaeoseismology. The collapses of the Selinunte temples (southwestern Sicily): two earthquakes identified. Bull Seismol Soc Am 92: 2961–2982

Guidoboni E, Ferrari G, Mariotti D, Comastri A, Tarabus G, Sgattoni G, Valensise G (2018a) The new version of the catalogue of strong earthquakes in Italy and the extended Mediterranean area (CFTI5Med): a modern tool for predicting future ground shaking based on historical seismology observations, 36th General Assembly of the European Seismological Commission, S02 - Seismological e Infrastructures, La Valletta, Malta

Guidoboni E, Ferrari G, Mariotti D, Comastri A, Tarabusi G, Valensise G (2018b) CFTI 5Med, Catalogue of Strong Earthquakes in Italy from 461 BC. to 2000 and in the Mediterranean area, from 760 BC to 1500, An Advanced Laboratory of Historical Seismology. http://storing.ingv.it/cfti/cfti5/

Guidoboni E, Ferrari G, Tarabusi G, Sgattoni G et al (2019) CFTI5Med. The new release of the catalogue of strong earthquakes in Italy and in the Mediterranean area. Nature 6:80. https://www.nature.com/articles/s41597-019-0091-9

Helly B, Rideaud A (2016) Ten years of archaeoseismological research in Armenia (2004–2013). Quat Int 395:216–232. https://doi.org/10.1016/J.QUAINT.2015.08.087

Hinojosa HR (2023) The importance of assessing the geological site effects of ancient earthquakes from the archaeoseismological point of view. Eng 4(1):719–737. https://doi.org/10.3390/eng4010043

Hinzen KG (2005) Archaeoseismological case study of the Main building of Kerkrade-Holzkuil. In: Tichelman G (ed) Het villacomplex Kerkrade-Holzkuil. ADC Archeoprojecten, Amersfoort, pp 407–420. ADC-Rapport 155

Hinzen KG (2011) Archäoseismologie: auf der Suche nach Spuren vorinstrumenteller Erdbeben; [533. Sitzung vom 9. Februar 2011 in Düsseldorf], Paderborn; München; Wien; Zürich: Schöningh

Hinzen KG, Schreiber S, Fleischer C et al (2013) Archeoseismic study of damage in Roman and medieval structures in the center of Cologne, Germany. J Seismol 17:399–424. https://doi.org/10.1007/s10950-012-9327-2

Hinzen KG, Montabert A (2017) Rectangular blocks vs polygonal walls in Archaeoseismology. Ann Geophys 60(4):S0443. [Internet]. Jul. 13 [cited 2022Jul.29], Available from: https://www.annalsofgeophysics.eu/index.php/annals/article/view/7127

Hinzen KG, Maran J, Hinojosa-Prieto H, Damm-Meinhardt U, Reamer SK, Tzislakis J, Kemna K, Schweppe G, Fleischer C, Demakopoulou K (2018) Reassessing the Mycenaean earthquake hypothesis: results of the HERACLES project from Tiryns and Midea, Greece. Bull Seismol Soc Am 108(3A):1046–1070. https://doi.org/10.1785/0120170348

Hinzen KG, Meghraoui M, Bahrouni N et al (2021) Archaeoseismological study of the Cherichira aqueduct bridge, Kairouan. Tunisia Med Geosc Rev 3(2021):403–430. https://doi.org/10.1007/s42990-021-00062-9

Jusseret S (2014) Contextualising the birth of Mediterranean archaeoseismology. Antiquity 88:964–974

Jusseret S, Langohr C, Sintubin M (2013) Tracking earthquake archaeological evidence in late Minoan IIIB

(~1300–1200 B.C.) Crete (Greece): a proof of concept. Bull Seismol Soc Am 103(6):3026–3043
Kaner S, Habu J, Matsui A (2011) Rescuing archaeology affected by the Japanese earthquake and tsunami. Antiquity 85(329) Available online: http://www.antiquity.ac.uk/projgall/kaner329/
Katsonopoulou D (2005) The earthquake of 373 BC. Literary and archaeological evidence. In: Katsonopoulou D, Soter S, Koukouvelas I (eds) Ancient Helike and Aigialeia, archaeological sites in geologically active regions, Helike III. The Helike Society, Athens, pp 15–32
Katsonopoulou D, Soter S, Koukouvelas I (2003) Evidence of earthquakes in Helike from the Early Bronze Age to Roman Times. In: 4th symposium on archaeometry, Athens
Kázmér M, Győri E (2020) Millennial record of earthquakes in the Carpathian-Pannonian region: historical and Archaeoseismology. Hung Hist Rev 9(2): 284–301. https://www.jstor.org/stable/26984122
Kázmér M, Al-Tawalbeh M, Győri E, Laszlovszky J, Gaidzik K (2021) Destruction of the Royal Town at Visegrád, Hungary: historical evidence and Archaeoseismology of the A.D. 1541 earthquake at the proposed Danube dam site. Seismol Res Lett 92(5):3202–3214. https://doi.org/10.1785/0220210058
Korzhenkov A, Mazor E (2014) Archaeoseismological damage patterns at the ancient ruins at Rehovot-ba-negev, Israel. Ernst Wasmuth, Tübingen
Marco S (2008) Recognition of earthquakerelated damage in archaeological sites: examples from the Dead Sea fault zone. Tectonophysics 453:148–156. https://doi.org/10.1016/j.tecto.2007.04.011
Monaco C, Tortorici L (2004) Faulting and effects of earthquakes on Minoan archaeological sites in Crete (Greece). Tectonophysics 382(1–2):103–116. https://doi.org/10.1016/j.tecto.2003.12.006
Nakamura F (2012) Memory in the debris: the 3/11 great East Japan earthquake and tsunami. Anthropol Today 28(3):20–23. https://doi.org/10.1111/j.1467-8322.2012.00874.x
Niemi TM (2009) Paleoseismology and archaeoseismology of sites in Aqaba and Petra. Field guidebook (PDF), Geological Survey of Israel, Jordan, pp 119–124. archived from the original (PDF) on 2013-09-29
Pecchioli L, Prihatmaji YP (2023) Sustainability of Indonesian wooden culture towards damage risk reduction, Restauro Archeologico (RA), n.2/2022, Firenze University Press (August). https://oaj.fupress.net/index.php/ra/index
Pecchioli L (2023) Disaster mitigation in Southeast Asia: sustainability of Indonesian wooden culture towards damage risk reduction. In: ICOMOS - ScSy21, Scientific symposium 2021 - poster session
Pecchioli L, Cangi G (2021) Interpretazione degli effetti sismici e analisi dinamica dei collassi in Ostia Antica. Restauro Archeologico 29(1):80–93. https://doi.org/10.13128/rar-11338
Pecchioli L, Cangi G, Marra F (2018) Evidence of seismic damages on ancient Roman buildings at Ostia: an arch mechanics approach. J Archaeol Sci Rep 21:117–127
Pecchioli L, Panzera F, Poggi V (eds) (2020) Cultural heritage and earthquakes: bridging the gap between geophysics, archaeoseismology and engineering. J Seismol 24. https://doi.org/10.1007/s10950-020-09936-1
Pecchioli L, Betti M, Pintucchi B (2022) Archaeoseismological project in Ostia tracking evidence on seismic damage by nonlinear numerical simulations. Develop Built Environ., Elsevier (available online 11.03.2022, 100073), In Press, Journal Pre-proof. https://doi.org/10.1016/j.dibe.2022.100073)
Pirazzoli P, Laborel J, Stiros S (1996) Earthquake clustering in the eastern Mediterranean during historical times. J Geophys Res 101:6083–6097
Poursoulis G, Dalongeville R, Helly B (2000) Destruction des édifices minoens et sismicité récurrente en Crète (Grèce) /recurrent seismicity and the destruction of Minoan constructions in Crete (Greece). Géomorphologie Relief Processus Environnement 6(4):253–265. https://doi.org/10.3406/MORFO.2000.1070
Poursoulis G, Levret A, Lambert N, Rideaud A, Helly B (2006) The 1708 Manosque earthquake (France): a reading of its archaeological traces as a contribution to estimate the effects on buildings. J Seismol 10(4): 511–526. https://doi.org/10.1007/S10950-006-9034-Y
Rodríguez-Pascua MA, Silva PG, Garduño-Monroy VH, Pérez-López R, Israde-Alcántara I, Giner-Robles JL, Bischoff JL, Calvo JP (2010) Ancient earthquakes from archaeoseismic evidence during the Visigothic and Islamic periods in the archaeological site of “Tolmo de Minateda (SE Spain)”. In: Sintubin M, Stewart IS, Niemi TM, Altunel E (eds) Ancient earthquakes. Geological Society of America, Boulder
Rodríguez-Pascua MA, Pérez-López R, Giner-Robles JL, Silva PG, Garduño-Monroy VH, Reicherter K (2011) A comprehensive classification of Earthquake Archaeological Effects (EAE) in archaeoseismology: application to ancient remains of Roman and Mesoamerican cultures. Quat Int 242(1):20–30., ISSN 1040-6182. https://doi.org/10.1016/j.quaint.2011.04.044
Russell Kenneth W (1980) The earthquake of May 19, A. D. 363. Bull Am Sch Oriental Res 238:47–64. https://doi.org/10.2307/1356515. JSTOR 1356515
Sintubin M (2011) Archaeoseismology: past, present and future. Quat Int 242(1):4–10. https://doi.org/10.1016/j.quaint.2011.03.056
Sintubin M (2013) Archaeoseismology. In: Beer M, Kougioumtzoglou AI, Patelli E, Au S-KI (eds) Encyclopedia of earthquake engineering. Springer, Berlin, Heidelberg. https://doi.org/10.1007/978-3-642-36197-5_29-2
Sintubin M, Stewart IS, Niemi TM, Altunel E (eds) (2010) Ancient earthquakes. Geol Soc Am Spec Pap 471:279
Stiros S (1996a) In: Stiros S, Jones RE (eds) Archaeoseismology. Institute of Geology & Mineal Exploration

Stiros S (1996b) Identification of earthquakes from archaeological data: methodology, criteria and limitations. In: Stiros S, Jones R (eds) Archaeoseismology, fitch laboratory occasional paper 7. Oxford, Athens, pp 129–152

Stiros S (2001) The AD 365 Crete earthquake and possible seismic clustering during the fourth to sixth centuries AD in the eastern Mediterranean: a review of historical and archaeological data. J Struct Geol 23:545–562

Stiros S (2021) Impacts of the 2020 Samos earthquake on the modeling of ancient seismic events. Turk J Earth Sci 30. https://doi.org/10.3906/yer-2107-7

Stiros S (2022a) Earthquakes, tsunamis, dried harbours and seismic coastal uplift: evidence from the eastern Mediterranean. Mediterr Geosci Rev 4:1–10. https://doi.org/10.1007/s42990-022-00079-8

Stiros S (2022b) The AD 365 Alexandria (Egypt) Tsunami. Revisited. https://doi.org/10.1007/978-3-030-73026-0_4

Valensise G, Guidoboni E, Iervolino I (2021) What is an exceptional earthquake? Seismol Res Lett 92:3309. https://doi.org/10.1785/0220210190

MIX
Papier aus verantwortungsvollen Quellen
Paper from responsible sources
FSC® C105338

If you have any concerns about our products,
you can contact us on
ProductSafety@springernature.com

In case Publisher is established outside the EU,
the EU authorized representative is:
Springer Nature Customer Service Center GmbH
Europaplatz 3, 69115 Heidelberg, Germany

Printed by Libri Plureos GmbH
in Hamburg, Germany